# Advance Praise for *Divergent Mind*

"Every woman who struggles to flourish in our one-size-fits-all society should read this book."

—Susan Cain, *New York Times* bestselling author of *Quiet: The Power of Introverts in a World That Can't Stop Talking*

"*Divergent Mind* explores powerful and positive ways to understand our fundamental differences. Its focus on women is much needed and timely. The book de-pathologizes those of us who most profoundly and intensely think and feel the world around us. I found it to be eye-opening and healing."

—Mary Pipher, *New York Times* bestselling author of *Women Rowing North* and *Reviving Ophelia*

"This book is powerful, much needed for our times, and Jenara Nerenberg offers a unique blend of personal, scientific, and societal analysis. *Divergent Mind* is really for all women, giving them the chance to understand each other's invisible differences and gifts."

—Elaine Aron, PhD, bestselling author of *The Highly Sensitive Person: How to Thrive When the World Overwhelms You*

"American psychiatry has increasingly taught our society to think that people can be divided into two categories: those who are 'normal' and those who are 'not normal.' In *Divergent Mind*, Jenara Nerenberg powerfully writes of an urgent need to scrap that way of thinking, and replace it with a societal appreciation for the many gifts that people with 'divergent minds' bring to our world. Hers is a clarion call for change."

—Robert Whitaker, *New York Times* bestselling author of *Anatomy of an Epidemic* and *Mad in America*

"In this enormously important book, Jenara Nerenberg brings to light the history and lived experiences of a 'lost generation' of neurodivergent women whose wisdom about coping in a world not built for them has been occluded behind a fog of changing diagnostic labels and overly male-centric research. *Divergent Mind* is a signpost to a happier future in which the diversity of human wiring is celebrated while barriers to access are eliminated."

—Steve Silberman, *New York Times* bestselling author of *NeuroTribes: The Legacy of Autism and the Future of Neurodiversity*

# Divergent Mind

# Divergent Mind

## Thriving in a World That Wasn't Designed for You

### JENARA NERENBERG

HarperOne
*An Imprint of* HarperCollins*Publishers*

HarperOne

HarperCollins books may be purchased for educational, business, or sales promotional use. For information, please email the Special Markets Department at SPsales@harpercollins.com.

FIRST EDITION

*Designed by Paula Russell Szafranski*

Library of Congress Cataloging-in-Publication Data

Names: Nerenberg, Jenara, author.
Title: Divergent mind : thriving in a world that wasn't designed for you / Jenara Nerenberg.
Description: First edition. | New York, NY : HarperOne, 2019 | Includes bibliographical references.
Identifiers: LCCN 2019033534 (print) | LCCN 2019033535 (ebook) | ISBN 9780062876799 (hardcover) | ISBN 9780062876805 (paperback) | ISBN 9780062876812 (ebook)
Subjects: LCSH: Women—Mental health. | Neuropsychology. | Brain—Sex differences. | Sensitivity (Personality trait)
Classification: LCC RC451.4.W6 N47 2020 (print) | LCC RC451.4.W6 (ebook) | DDC 616.890082—dc23
LC record available at https://lccn.loc.gov/2019033534
LC ebook record available at https://lccn.loc.gov/2019033535

20 21 22 23 24   LSC   10 9 8 7 6 5 4 3 2 1

*For my family*

# di·ver·gent

/dəˈvərjənt,dīˈvərjənt/

*adjective*

1. tending to be different or develop in different directions; "divergent interpretations"

Synonyms: differing, varying, different, dissimilar, unlike, unalike, disparate, contrasting

# Contents

# Introduction

When I moved back to California after six years of reporting from Asia, my daughter, who was two and a half years old at the time, cried, "Mom, you're just running around and around and around!" "Oh my gosh," I thought, "she *sees* me; I have no idea what the hell I'm doing, and now she's found it out." That same year, the National Institutes of Health announced $10.1 million in grants to counteract gender bias in research. I wish I had known at the time, because I would have volunteered myself as a subject. I was depressed, confused, anxious, tired, and plagued with a persistent feeling of inadequacy and the feeling that somehow I wasn't myself. I would drop my daughter off at her new preschool and feel as though I was wearing a mask around all the other parents, for fear that they would discover how inept I was. Meanwhile, increasing tensions at home with my husband made it even harder to manage the mess of feelings—and the literal mess of dishes and laundry.

I was stumped—here I was, a graduate of the Harvard School of Public Health and UC Berkeley, with reporting for CNN, *Fast Company*, Healthline, and elsewhere under my belt, and yet I couldn't for the life of me figure out how to make and stick to a decent schedule, stay on top of household duties, or hold logistical conversations with my husband.

Meanwhile, between 2013 and 2016 a slew of articles were published in major news outlets about how adult women were being overlooked in research about "attention-deficit/hyperactivity disorder" (ADHD) and autism. Writer and ADHD activist Maria Yagoda published a story in *The Atlantic* about being a student at Yale who struggled with a broad range of executive functioning, from cleaning to losing items and money, to tracking what time to be where. People didn't believe she could have ADHD because she was "so smart." Spectrum published a similar story about girls and women with autism; a young woman named Maya was profiled who was thought to struggle with severe anxiety and other challenges until she was finally recognized as being autistic.

I think Facebook was listening to my conversations with family members and therapists because suddenly articles and books started popping up in my news feed about women with ADHD and Asperger "syndrome" and about the highly sensitive person (HSP). I started seeing new research coming out about the mental health challenges of "high-achieving" women, and I was transfixed. Captivated. Utterly glued to what I was reading.

Because it turned out that I wasn't alone. Study after study indicated high rates of depression and anxiety among "successful" women, but other traits, like ADHD and autism, were beginning to surface as well. I had never thought of these before, but I couldn't deny that what I was reading was resonating with me. I was "sensitive," I liked talking about only a few select topics (people, psychology, and inner life—my "special interests"), logistics might as well have been an alien language, and this word *masking* kept jumping out at me, as describing an experience that I didn't realize or want to admit to.

So I begin this book by sharing a kind of confusion that plagued me at the time—a feeling of shock and dissonance but also of hope and relief. Could I be on some kind of sensory spectrum, like the autism spectrum? Or did I have ADHD? Both seemed likely. But I didn't seek out assessment, diagnosis, confirmation, or anything of the like. Instead, I turned to research, studies, news articles, and countless interviews and stories with women who sounded a lot like me.

## Masking

*Masking* refers to an unconscious or conscious effort to hide and cover one's own self from the world, as an attempt to accommodate others and coexist. Research and anecdotal evidence show that an extensive amount of masking and "passing" is going on among women and girls, primarily because of the way women are socialized. Girls and women have been taught from an early age to "blend in," according to researchers and the many women I interviewed for this book. Often, women hear the common refrain "Oh, she's just sensitive. That's how girls are." This is a sloppy, but widespread, oversight in our culture.

Masking claims many lives—and I don't necessarily mean that women literally commit suicide (although that can happen as well), but they may commit a kind of virtual suicide—leaving many women feeling empty, depressed, and anxious and robbing them of living according to their true selves. When society is not equipped to hold an accurate mirror up to you, you end up interpreting your reflection according to available lenses, structures, and terminology. But they're often wrong and misleading, or, worse, harmful.

My depth of curiosity, sensitivity, persistent wondering, and questioning—my insatiable hunger to know and understand—is not mirrored in the wider culture, even in academia. I am deeply curious about the inner lives of others and understanding them—which often looks like asking a ton of questions (good thing I became a journalist), but this is not how people make friends. It took me a long time to figure that out. So instead of accepting myself as curious, passionate, and inquisitive, I felt different and isolated. Slowly, I allowed my mannerisms and gestures to match those I witnessed around me and the messages I was getting—namely, don't ask too many personal questions, don't talk too much, don't deliver essay-length monologues on philosophical topics.

Over time, I changed; some of the change was likely because of natural maturation, but certainly some of it felt painful and necessary in order to adapt. I had to "stuff" a lot of my curiosity, and I turned to reading and to other independent ways to explore the expanse of the mind—meaning I spent less time interacting with other people and more time alone. Again, these are not bad things in and of themselves, but at the time, I was operating in a binary of "abnormal" and "normal" and thought those were my only options in order to coexist with the world. I had no knowledge yet of the wide diversity of ways that the brain is made up and ways that people interact. So without anything telling me otherwise, without any mirrors reflecting who I was, I masked and suppressed.

This is happening for women across the globe. In the past we were labeled "hysterical," but now we're "anxious." What many women don't know—and this includes the doctors and therapists with whom we interact—is that other mirrors are available to us that reflect previously hidden parts of ourselves.

It is said that the senses can be gateways to the soul, and I take that quite literally. Sight, sound, taste, touch, and smell correspond to either our mental health or our mental distress, depending on our sensitivities. Think of an onion, with its many layers: at the core of our being are our genes, biology, and childhood experiences but also our sensory makeup— that is, how our nervous system responds to and interacts with our sensory world, what delights us and what repels us. Over time, throughout our lives, all of these components interact, producing layers of emotions and resulting behaviors. When some of us end up in a therapist's or doctor's office with anxiety, depression, or autoimmune health challenges, our options are limited to talk therapy or medication because only these outer layers of emotion and behavior are probed. We have been going about our lives and professions thinking we know the full list of possible diagnostic criteria, but the senses have been left out; and thus a very core component of what makes people who they are goes completely untended.

Many people think of outdated, stereotyped images when they consider autism and ADHD, and it's important to remember that there is a spectrum of experiences. It's likely that these labels could apply to people in your own life—perhaps your boss, neighbor, friend, or family member, or even you. What I see as fundamentally missing from the conversation is a rallying point around diversity in how individuals process sensory input—*and specifically, recognizing a broad occurrence of heightened sensitivity.*

Some people may be fine with leaving this discussion at what has already been expertly explored by Elaine Aron in her 1996 book *The Highly Sensitive Person: How to Thrive When the World Overwhelms You.* But I'm not—I want psychology and psychiatry to take sensitivity concerns further, because of how they affect people's work, family life, education, economic opportunities, intimacy, and parenting. The public and professionals need to understand that people with sensory differences—such as autism and ADHD and a few other "neurodivergent" traits we will learn about in this book, such as sensory processing disorder (SPD), HSP, and synesthesia—experience heightened sensitivity across the board, and these differences affect nearly every aspect of their lives.

Many such neurodivergent women are suffering, as many times these traits occur along with anxiety and depression, especially if the underlying sensory differences go undiagnosed. The full scope of such differences—which often merit a diagnosis—is often unknown not only to friends and family but *also to the women themselves.* One woman I interviewed graduated from Columbia University but wasn't "diagnosed" until the age of twenty-eight. Likewise, a mom in California didn't realize her own autistic and ADHD traits until she was in her forties and her son was diagnosed; she recognized that his symptoms were ones that she had been dealing with her whole life. I realized my own neurodivergences at the age of thirty-two only because I started digging into the latest research.

An entire demographic of women is now being referred to as a "lost generation," because an extensive amount of depression and anxiety surface as a result of internal experiences that don't match up with what the world expects or how the world views such women—since they appear to function "normally" on the outside. This lack of awareness and understanding is largely due to neglect on the part of researchers because study samples often rely on streamlined populations of men; therefore, doctors, therapists, teachers, and police officers just don't know what a woman with ADHD, Asperger's, synesthesia, SPD, or high sensitivity might "look like" or how she might act. As a result, thousands of women have no name for their life experiences and feelings.

When I began encountering questions myself, I started digging deeper, fueled by wanting to have a name, a label, for my experience in the world—how I "show up," how my mind and body react to certain situations, and most of all why I have felt so bad about myself.

## Neurodiversity: The Game Changer

One day on a flight from South Korea to Nepal, I started imagining that there may be more people out there like me—and what if there were other ways of "being" in the world that didn't have names or labels yet, especially for women? I created the phrase *temperament rights* to capture this idea of one's temperament or neurological makeup being respected in the same way that we respect other core aspects of people, such as gender, sexuality, or ethnic identity. I started to imagine a world in which the richness of the human interior—what we call one's "inner life"—is acknowledged and respected with the same awareness of diversity that we see in terms of outer categories of identification such as race, culture, sexuality expression, and gender.

If courageous leaders and activists before me had rallied around the importance of recognizing these outer categories, couldn't we do the same for internal categories of identification? *Don't our inner lives deserve just as much attention as our outer lives?* Having an inner life and an internal emotional world is universal; it's something all of us possess. And something like ADHD or high sensitivity can show up in anyone—women and men, white folks and people of color, trans folks and cis folks.

Being the person I am, I rushed to my laptop and started searching around to see whether others were talking about this. It didn't take me long to find the term *neurodiversity*, which means recognizing and celebrating the diversity of brain make-ups instead of pathologizing some as "normal" and others as "abnormal."

What happened next is one of those moments in life that haunt me—in the best of ways. I had been noticing a striking man walking my street early in the morning each day, with his daughter beside him. I sensed a joy within him, a freedom, a walk that signaled openness and calm and groundedness. His ever-present smile was notable. I passed him almost every day while walking my own daughter to her school. And when I was at my computer months later in Asia and discovered neurodiversity for the first time, this same man's face popped up on the screen! It turns out that he, Nick Walker, is a notable neurodiversity author and scholar—and he lived just a few blocks away from me. At the same time, and thanks to Twitter's algorithms, I found the tweets of author and neurodiversity expert Steve Silberman and started to delve into what would come to define the next few years of my life—an exploration and investigation of neurodiversity.

## Sensitivity

Silberman's 2015 book *NeuroTribes: The Legacy of Autism and the Future of Neurodiversity* is a historical account that focuses largely on boys and men with the neurodivergence of autism, but I found myself leaning into the research about adult women and several neurodivergences that have high sensitivity in common. From my research, I discovered that the trait of sensitivity seems almost synonymous with developmental neurodivergences in adult women.

*Sensitivity* implies a certain heightened reaction to external stimuli—experiences, noise, chatter, others' emotional expression, sound, light, or other environmental changes. Sensitivity and high empathy are common experiences for many women, but some experience these qualities to more severe degrees, and they remain unaware that they can be hallmarks of Asperger's, ADHD, HSP, and other traits. (Note: I use the words *woman* and *female* interchangeably in this book because of their usage in academic research, but the experience of sensitivity and a woman's experience generally is clearly genderless, nonbinary, and equally applicable to trans women and cis women.)

Elaine Aron's use of the term *high sensitivity* in her book *The Highly Sensitive Person* refers to a person with a characteristic depth of processing of external information—a person with *sensory processing sensitivity* (SPS), which is the scientific term for HSP. For someone with Asperger's, sensitivity might imply a sense of being overwhelmed when overstimulated.

And for someone with ADHD, a common but unknown feature is a sensitivity to one's own emotions and the regulation of them. For the person with SPD certain smells or textures heighten their reactions. And for the person with synesthesia (a *synesthete*), the presence of suffering or strong emotions in others can overwhelm them, an aspect of synesthesia called "mirror touch." It is interesting to note that all five of these neurodivergences—HSP, ADHD, autism, SPD, and synesthesia—often imply some version of "melting down" emotionally—adult tantrums, quick-appearing migraines, outbursts of anger—because of sensory overload.

Once we understand sensitivity, and its connection to neurodiversity, sensitive women no longer have to walk around with a hidden secret about what they know they feel and experience every day—taking in vast amounts of information about one's environment, including the people in it, and *somatically processing all of that input*. The science has finally caught up with our real, lived experience, and we no longer need to hide in a closet for fear of being deemed "crazy," overemotional, or not academic enough.

*Divergent Mind* explores specifics about five neurodivergences that have sensitivity at their core—HSP, ADHD, SPD, autism/Asperger's, and synesthesia—and how new understandings and insights can be applied to daily life and society as a whole. We will dive into the worlds of women who have spent their lives masking—*without knowing it*. Because of the way women are socialized to "fit in" and pick up on social cues, underlying traits of autism or ADHD or other neurological makeups *essentially get missed*.

So again, enter neurodiversity—the understanding whereby mental differences are viewed simply as they are and not judged as better or worse, normal or abnormal. As a society, we need a shift in thought that applies to all neurological makeups—including more well-known ones such as bipolar "disorder" and schizophrenia—but this book focuses specifically on five "sensory processing differences," with sensitivity at their core, that would typically be classified as being related to "developmental" differences (with the exception of HSP). Neurodiversity is a paradigm shift that empowers women to come forward, be seen, better understand themselves, and proudly claim their identities.

*Divergent Mind* also highlights the pressing need for our definitions of "mental health," "disorder," and "mental illness" to evolve. For example, is ADHD a "disorder," or is it simply one form the human brain takes in our species as part of a natural array of human brain diversity—much as biodiversity implies a variety of plants, colors, and fauna in an ecosystem? Other questions we explore include How do we make space for the variety of human brains and sensory makeups we see? What happens when we stop pathologizing difference? We'll see that creativity, innovation, and human flourishing often result. By adopting neurodiversity thinking, we can begin to see brain difference and sensory difference as any other difference we acknowledge and celebrate.

Furthermore, how does knowing that neurodivergent people make up *at least* 20 percent of the population begin to shift our concept of "normal," "disordered," or "mentally ill"? Perhaps we are really talking about *humanity* as a whole rather than a set of "neurotypical" versus neurodivergent individuals. Given that so many neurodivergent people go undiagnosed, we may be looking at an entirely different concept of what it means to be human.

Such a shift in understanding could help thousands of women around the world living with undiagnosed or misdiagnosed neurodivergences avoid years of unnecessary comorbidities such as depression, anxiety, shame, guilt, low self-esteem, and distorted self-image. Neurodiversity, when embraced, can dramatically improve all aspects of life.

## The Conversation

What is considered pathology is largely a construct and product of the times. Mental health specialists spend their careers carving out the precise parameters around certain "diagnoses," and when two or more diagnoses start to overlap or run up against each other, people get territorial and defensive and protective. This may sound shocking or ridiculous, but it is true. So it is imperative that the language and vocabulary of neurodiversity—the understanding that there is a natural array of human brain makeups—begin to seep not only into the medical and psychiatric canon, but also into the everyday colloquial language of the public.

We must ask, *Why does the way you pay attention determine your work prospects and life satisfaction?* If you, like me, pay attention either in spurts or overwhelmingly on one thing (called "hyperfocusing"), then your teachers or bosses may start to view you as if you don't align with the "norm." Thus, you unknowingly begin to edit and adapt—to mask—for survival. This begins a repeating cycle of censoring, attempting to fit in, and overall altering your performance of your "self" in the world, leading to depression, anxiety, burnout, or worse. (The correlation between neurodivergence and having suicidal thoughts is staggering.) You're also more likely to be fired from jobs than people who are neurotypical, and thus you may struggle financially. All the while, however, you also know you've been able to perform well in particular areas, especially in unconventional working environments. This all paints a confusing picture, and you may get to the point of asking yourself, like I did, "What the hell is going on?" Your confidence and self-esteem plummet, and you begin to question many of your oldest experiences and frames for understanding yourself—and society.

At a gathering hosted by Krista Tippett of the public radio show *On Being*, a group of women and I were sharing our stories, and one woman sitting across from me began to cry because she had lost her daughter to suicide a few years before and the intimate discussions were helping her embrace her vulnerable feelings. As she started to describe her daughter, I told her some of what I had been learning about sensory differences, and she cried, "I didn't know! I didn't know!" For her daughter and for many other women, what looks like "bad" or "failing" behavior is sometimes a response to overwhelming sensory input, but the only thing that is noticed is the outer behavior. It is crucial that we question anxious, depressed, or "inappropriate" behavior through these new understandings and research about sensory-related mental health challenges.

As I interviewed women for this book, I heard over and over how life-changing it has been for them to discover simple tools and techniques to help themselves regulate emotions and resulting behaviors. In this book you will learn about not only such tools, but also what is happening in the world of architecture, design, virtual reality, and more so that women like us can breathe and walk around feeling peace and a sense of place and happiness. We are creating a brand-new ecosystem so that we can thrive. Creative, entrepreneurial projects and collaborations are popping up across the globe. Women like us are defining new identities, new cultures, new ways of communicating and interacting—ways that suit our temperaments.

And guess what? It's not just for us. As we innovate, our ways begin to spill over into the general population. The way we see, describe, interpret, design, educate, or collaborate will begin to be viewed as more than just "different" and slowly but surely will become more of a norm whereby the world that previously felt foreign and isolating to us finally begins to feel like home—a skin we are finally comfortable in. So *Divergent Mind* is for neurodivergent folks, but also for the friends, educators, parents, doctors, partners, and work colleagues of those folks.

## A Note on Language

*Social awkwardness* is a term that has enjoyed some buzz recently, but I think some caution is needed here. If we call attention to neurodivergence as a form of social awkwardness, then we are further enforcing the idea of "normal" and "abnormal" and the dominance of the neurotypical status quo. If we wish to move to a language and framework of friendly neurodiversity, the term "socially awkward" must be removed. *We are all different flavors of human.* There is no one "correct," "right," or "standard" way to be. There are tendencies, yes, and for that we have the label "neurotypical," but as research increases on the individual variations of our brains and temperaments, I believe each particular strand of brain makeup will be viewed simply as that—the same way each color is viewed for what it is, with none deemed more "normal" than the next.

That said, it is important to understand some of the basic terms used in this book. The term *neurodiversity* was first put forth in the late 1990s by the Australian sociologist Judy Singer to capture the array of brain makeups found in the human species. The need to pathologize some brains over others would be less necessary under such an umbrella term, and the inherent diversity would simply acknowledge natural differences. Steve Silberman popularized the term in his groundbreaking book *NeuroTribes*, which focuses on the history of autism and the potential future of thinking within, and acting on, a neurodiversity paradigm. Autistic scholar and educator Nick Walker proved to be a close ally of and source of support to Silberman and other proponents of neurodiversity. Walker's popular blog *Neurocosmopolitanism* is a definitive source for vocabulary and definitions for these and other terms.

If neurodiversity is the umbrella term—a fact of the human species—then *neurodivergent* is the term to be applied to individuals. If a person has been labeled or identifies as ADHD, autistic, bipolar, dyslexic, or another "diagnosis," that person is said to be neurodivergent. Within a neurodiversity framework, any person possessing or identifying with a "mental illness" or "developmental disorder" may be considered neurodivergent. (Often, people have no label or diagnosis but strongly sense there is something different about them.)

A neurodivergent person possesses what is called a *neurodivergence*—the cluster of behaviors or signs that led to that person's label or diagnosis. For example, you may identify as HSP or as a synesthete, so you would refer to those as your neurodivergences. This shift in language and narrative is empowering and less pathologizing.

A person who does not have any neurodivergence is *neu-*

*rotypical.* Neurodivergent people often use this term as a way of differentiating styles of communication, expectations, and work that come up in their relationships with such people.

The organizing and advocacy that has evolved to support a neurodiversity framework and how that might be applied at school, work, and home within families is known as the *neurodiversity movement.* Many compare this movement to the kinds of changes and understandings accomplished by the civil rights movement, the women's rights movement, and the gay rights movement. They share a common language, though admittedly the context and conditions vary.

The arrival of a neurodiversity framework, situated within the history of psychology and psychiatry, thus poses critical questions. With the increasing "rise" of modern neurodivergences such as ADHD and autism, for example, how do we choose to respond? How do we frame such neurodivergence to begin with? Do we learn from history and take a wider perspective that incorporates a societal and contextual lens, or do we focus on individuals? Are these terms simply the latest historical expressions of where the individual's natural orientation rubs up against the expectations of society? And what do such rising rates signal to us about how society exists now?

## Onward

Neurodiversity thinking is taking a firm hold within organizations, schools, corporations, design agencies, therapy practices, individual families, top tech companies, architecture companies, and Ivy League universities. A variety of disciplines are beginning to cross-pollinate ideas, and often those people at the helm of such collaborations are themselves neurodivergent. Why? And why now?

As we begin to shift away from previously held conceptions about gender roles, identity, sexual expression, race, and ethnicity, we are also beginning to be free from definitions and expectations *about how to think and act in the world.* The ADHD person blends her or his variety of interests into a new academic research center at a top university. The autistic teacher encourages more movement in the classroom to allow for fuller expression of hand movements, or "stimming," to help those with sensory sensitivities. The synesthete leverages the science of virtual reality to inform chord progression on a new song. The HSP architect designs new spaces conducive not only to sensitive people but to the overall mental health, calm, and well-being of the general public. The SPD fashion designer designs more comfortable clothing for kids and their parents.

It is essential, then, to understand how sensitivity—which is at the core of so many modern-day neurodivergences—operates and shows up and how to honor it, along with the neurodiversity framework, to generate widespread healing, especially for women. The sensory aspects that dominate the descriptions of so many labels today simply cannot be ignored and tell us important information about how well society is functioning and what is not working for a sizable chunk of the population.

## This Book

My training as a journalist and my lived experience inform this book. As the founder of The Neurodiversity Project, a series of community gatherings and author events for progressive research in medicine, design, the arts, and psychology, I have learned much from integrating research, personal experience, and observing healing in community. At this point in time, we have extensive research on individual traits—we've all read about ADHD or autism in the news—but how does the picture look as a whole, for neurodiversity and especially for women? I believe that years of suffering could be avoided if women were better studied and the emerging research were more widely known.

I wrote this book primarily to empower women who have a deep sense that they are "different" from the "norm" but would never in a million years think they "have" ADHD, Asperger's, or some other neurodivergent trait. How could they, since the research has barely focused on them, that is, on women? This book is for those soon-to-be-known-as neurodivergent women and their families, friends, colleagues, and all the other people surrounding them. Being able to finally give a name to an experience is incredibly healing and liberating. Once these women realize who they are, and once the world embraces what makes them unique, perhaps we'll finally be able to utilize their strengths—because it's not just about alleviating suffering, but the opportunity to improve our society as a whole.

We've heard from related books in sociology and neuroscience and the mental health and wellness categories. Susan Cain's 2012 book *Quiet: The Power of Introverts in a World that Can't Stop Talking* empowered an introversion revolution, and Steve Silberman's *NeuroTribes* traced the history of autism and the potential of the neurodiversity framework. Brené Brown's 2012 *Daring Greatly: How the Courage to be Vulnerable Transforms the Way We Live, Love, Parent, and Lead* and other books

showed the transformative power of revealing our inner vulner-
abilities, and Elaine Aron's *The Highly Sensitive Person* (1996)
gave us a nonpathologizing language of sensitivity for the first
time.

*Divergent Mind* takes up the task of exposing the hidden
inner worlds of those with sensory differences and challenging
the world at large to listen and shift course. It takes readers on
a journey of imagining what the world would be like if neurodivergences of all kinds were accepted and embraced. Would
the "crazy" aunt produce magnificent works of art for sale
while living in the extra bedroom of her sister's house? Would
the ADHD woman at work finally be able to thrive if her gifts
for flexibility and troubleshooting were embraced by managers
and colleagues and she became the go-to person in times of
crisis? Would the autistic woman be respected as an international public speaker because of her cutting-edge insights on
her "special interest," much as climate change student activist
Greta Thunberg has demonstrated?

As a society we are crumbling by staying stuck in an outdated, factory-inspired mode of operating that simply does
not work with the large demographic of people we call neurodivergent. This book finally unravels the reasons why so many
women are dealing with shame, guilt, and poor self-image in
the face of a neurotypical society that does not make space for
them. Everything shifts when these unseen sensory differences
are finally recognized and spoken aloud. Once women realize
the truth, they can leap light-years ahead in their lives.

# Part I

# INNER WORLDS

## Chapter 1

# The Female Mind Throughout History

Words, language, definitions, and framing all act as power conductors—they let in meaning, set boundaries, keep unwanted implications out, and generally empower or disempower. When we think about our choice and use of the phrase *mental illness*, for example, we have to stop ourselves and ask some questions. Who came up with this term and when? Was it a man? A scientist, pastor, plumber, farmer? What else was happening at the time the words were being employed? Was there slavery? Child marriage? Lobotomies?

I urge you to take questions of language seriously—not because you need to become a historical or linguistic expert, but because seriously exploring and answering such questions helps generate broader societal shifts. If your definition of your mind has been "locked" in an outdated viewpoint, or if how doctors or researchers view the mind is stunted, then increased questioning will begin to chisel at and loosen the very critical and often private world of the emotional interior.

The worlds of medicine and psychiatry have been "plagued" with words of weight that greatly affect the lives of women. We take the words and definitions for granted, unaware of their histories and unaware of the meanings in which we swim, until at some point in our lives we begin to pull (or push) back.

What women are subjected to—both in practice and in viewpoints—always reflects the broader sociocultural dynamics at play. During slavery in the United States, for instance, people began naming "slave diseases"—when slaves showed signs of unhappiness and wanting their freedom. In other eras, as women began working outside the home and gaining more freedom, doctors advised them to stay home for fear of injuring their reproductive organs. Homosexuality was deemed a mental disease until 1973. There is an interplay, an ontological dynamic between self-perception and societal structure.

In the 1400s, for example, the common notion of madness was that the devil and evil spirits had taken possession of the human mind. This belief contributed to many women being viewed as "witches" and then killed, especially if they were considered to be countercultural or irreligious, as Denise Russell notes in her book *Women, Madness, and Medicine* (1995). By the 1700s madness became a notion of human weakness rather than spirit possession, and into the nineteenth century female "hysteria" started to become a common reference instead.

Medicine and psychiatry have always toggled between being

viewed as a "divine" or a "scientific" practice; in fact, psychiatry partially originated in the disciplines of obstetrics and gynecology, where it was seen as "rightly" housed within the study of women's issues. In the late 1800s many male doctors asserted that in some cases the female clitoris should be removed because hysteria was caused by masturbation.

Hysteria was described as a "disease" of "uncontrollable emotion" that predominantly affected women, and "sexual emotions" encouraged it. Others described hysteria as a physical disorder that caused women to be morally perverted or to suffer from "paralysis of the will." Whatever feelings and bodily complaints women described at the time were generally not believed by doctors, who discarded the complaints as lies and caused by laziness. The common cure for hysteria was thought to be marriage. Henry Maudsley, a prominent British physician in the 1800s, claimed that menstruation could lead to mental instability and mania. In 1875, another British physician, Andrew Wynter, believed that insanity was passed down to children from the mother.

In 1895 Sigmund Freud published *Studies on Hysteria*, which concluded that early sexual trauma was responsible for adult female hysteria. Just two years later he switched his viewpoint because widespread reports of childhood sexual abuse led him to believe that his theory must not be entirely accurate. The pressure to pathologize and medicalize, using the language of disease, put weight on Freud to conform, or he would lose his stature in the medical community. No one could imagine that hysteria wasn't a disease—no one saw the connection between the way women were treated and their emotional lives.

In her 1985 book *The Female Malady: Women, Madness, and English Culture, 1830–1980*, Elaine Showalter traces the history of psychiatry from its association with gynecology to the pervasive overrepresentation of women in mental hospitals, and onward to the 1960s when studies concluded that mental illness was found more often in women than men. Showalter points to the "dual images of female insanity—madness as one of the wrongs of woman; madness as the essential feminine nature unveiling itself before scientific male rationality." Calling to mind the writers Sylvia Plath, Virginia Woolf, and Anne Sexton, Showalter continues, "Biographies and letters of gifted women who suffered mental breakdowns have suggested that madness is the price women artists have had to pay for the exercise of their creativity in a male-dominated culture."

Psychology and psychiatry as we know them today had to fight hard to become respected disciplines, and they didn't become so until they were firmly accepted and entrenched within the practice of medicine and science more broadly. As ideas regarding madness, insanity, and disorder were slowly evolving, many practitioners were viewed as "quacks." Scientists from other fields developed tools to examine body parts and physical disease, and the same approaches were attempted with the brain and human behavior. The development of psychology and psychiatry thus evolved as science as a whole evolved.

But human behavior, desires, beliefs, and thoughts are harder to probe, so the two fields took longer to develop. North America's first medical college was established in 1765 in Philadelphia, signaling the arrival of medicine and medical studies as "scientific." But doctors were still treating "insane" patients through "bleeding," a way of purging and fixing so-called irregular blood circulation. "Drowning therapy" was also popular at the time, as was making patients extremely dizzy by spinning them around and around.

The increased medicalization of challenges affecting the brain and behavior meant that less and less attention was being given to how broader sociological and historical contexts would affect people's emotions and mental states. During the late 1700s there had been a short-lived wave of what was called "moral treatment," where people who were deemed insane were treated slightly better and lived in quarters with gardens and areas devoted to the practice of art. A man by the name of Philippe Pinel in France was the primary force behind this approach, and soon after the Quakers in North America were experimenting with similar approaches.

But when these centers started to be influenced by medical professionals, the focus on moral treatment lessened, and a hybrid approach combining medicine and moral treatment began to take hold. Physicians began to manage these centers and in turn medicalize interactions with patients as opposed to relying on more gentle, human dynamics that brought out "balance" in the individual. As Robert Whitaker writes in *Mad in America: Bad Science, Bad Medicine, and the Enduring Mistreatment of the Mentally Ill*:

> As physicians gained control of the asylums, they also constructed a new explanation for the success of moral treatment—one that put it back into the realm of a physical disorder. Pinel's "non-organic" theory would not do. If it were not a physical ailment, then doctors would not have a special claim for treating the insane. Organizing activities, treating people with kindness, drawing warm

baths for the ill—these were not tasks that required the special skills of a physician.

Beyond the influence of medicine, larger contextual forces were underestimated as well. Much of what became known as "madness," especially in women, cemented around the time of the acceleration of capitalism in Western Europe. If context determines "normal" and "abnormal," then it makes sense to give greater attention to broader historical, economic, and societal forces at work. Showalter notes that since the eighteenth century, there were often literary references to a kind of "English malady": "The English have long regarded their country, with a mixture of complacency and sorrow, as the global headquarters of insanity." And as England's influence on the cultural development of the United States is vast, it's worth noting how notions of madness came about in contemporary US history.

Between the early 1800s and early 1900s, "The everyday psychopathology of the masses was a burgeoning and protean market, especially among the swelling ranks of the affluent; and doctors, armed with the authority of the microscope and the pharmacy, had seized it," explains Gary Greenberg in *The Book of Woe: The DSM and the Unmaking of Psychiatry*. At the same time, the idea that the transition to modernity was creating a rising incidence of "mental disorder" or "insanity" would occasionally crop up. Edward Jarvis, a physician in Massachusetts in 1872, began to speak and write on this, but it never gained much ground.

The demise of moral treatment also coincided with a wave of new immigrants leaving Europe for the United States. The switch from moral treatment to what people were swayed to believe was "hard science" was led by William Hammond, the

Surgeon General during the US Civil War, who claimed that insanity was definitely a brain disease. By the 1930s it became illegal for the "insane" to marry in order to prevent their procreation. About four thousand mentally ill patients were sterilized in the 1950s, a number similar to that of the 1920s. And as late as the early 1950s, around the same time as the first *Diagnostic and Statistical Manual of Mental Disorders* (the DSM) debuted, about ten thousand patients were still undergoing lobotomies for the treatment of mental illness.

The medicalization of "abnormal" behavior, which became the field we now know as psychiatry, is thus a history laced with layers and layers of distant players. The pursuit of medicine as a career was not always prestigious. Only as the study of medicine became accepted and entrenched in the scientific fields did its prestige rise. So to begin to discuss human behavior, abnormalities, and so-called disorders within a medical framework was a bit like colonists arriving on new land to make their mark—and a hefty profit. Academics, scientists, government officials, and business professionals got involved to essentially grab a piece of the market share. Greenberg writes:

> Surely the doctors who insisted that homosexuality was a disease were not all bigots or prudes. Nor are the doctors who today diagnose with Hoarding Disorder people who fill their homes with newspapers and empty pickle jars, but leave undiagnosed those who amass billions of dollars while other people starve, merely toadying to the wealthy. They don't mean to turn the suffering inflicted by our own peculiar institutions, the depression and anxiety spawned by the displacements of late capitalism and

postmodernity, into markets for a criminally avaricious pharmaceutical industry.

The history of the DSM thus reflects the history of politics and sociocultural dynamics. By the 1960s, the language of neurotransmitters, dopamine, and serotonin took hold. But, as Greenberg notes, "What seemed never to be in doubt as the doctors rushed from theory to theory was the idea that one brain chemical or another was the cause of mental suffering, just as one bacterium or another must be the cause of infection." There thus arose the pressure to further categorize and itemize every possible "disorder" and its criteria. As Greenberg notes, the third iteration of the DSM (DSM-III), which came out in 1980, "nearly [doubled] the number of mental disorders" and "vastly expanded the manual's scope, turning it into an entirely new psychopathology of everyday life." It was actually a bestseller and garnered the American Psychiatric Association an extensive amount of money. Its success was largely due to seemingly highly scientific categorizations of so-called mental disorders. But those categorizations are fluid and changeable: from slave drapetomania in the 1850s (the "mental illness" that caused slaves to try to escape slavery) to homosexuality in the 1950s to today's "internet use disorder." The DSM is more like a catalogue of current social ailments than scientific hardwired "diseases."

In *Women and Madness*, Phyllis Chesler writes of what she calls "psychiatric imperialism," whereby normal responses to trauma are methodically pathologized in science and medicine. At the time of the book's publication in 1972, few women were coming forward about gender biases in the study and practice of psychology. Chesler felt compelled to bring forward a conversation around gender, race, class, and medical ethics because "modern female psychology reflects a relatively powerless and deprived condition." Of sensitivity she writes: "Many intrinsically valuable female traits, such as intuitiveness or compassion, have probably been developed through default or patriarchal-imposed necessity, rather than through either biological predisposition or free choice. Female emotional 'talents' must be viewed in terms of the overall price exacted by sexism." Regardless of causation, of note here is that women's internal lives were barely acknowledged or considered.

A 1984 questionnaire from the National Institute of Mental Health indicated that close to one-third of Americans would experience "mental illness" during their lifetimes. And more than 20 percent of Americans reported symptoms that would land them a "diagnosis" based on the DSM. Given such numbers, Greenberg notes that "our inner lives are too important to leave in the hands of doctors: because they don't know as much about us as they claim, because a full account of human nature is beyond their ken."

In the later half of the 1980s, for example, ADHD became a hot topic and popular diagnosis, and a strand of kids caught the attention of a doctor named Joseph Biederman at Massachusetts General Hospital. These kids were precocious and prone to tantrums and extreme sadness. Biederman sensed that perhaps the onset of bipolar disorder was much earlier than previously thought and that these kids might be showing early warning signs of it. Features of ADHD and bipolar disorder in the DSM overlapped, but these kids didn't have the manic episodes characteristic of bipolar. The layers of criteria and nuanced symptoms listed in the DSM that are designed to differentiate one diagnosis from the next are fluid and relative, contributing to what Greenberg calls "ad-hoc diagnosing."

Throughout the early 2000s Biederman's newfound focus on childhood bipolar disorder resulted in a surge of medications being prescribed to children. According to Greenberg, the diagnosis of bipolar disorder among children in 2003 had increased by fortyfold over the previous decade. By 2005, antipsychotic drug use in kids increased 73 percent in just four years. In 2007, five hundred thousand children were on drugs that previously had been reserved for extreme cases.

This is an example of the consequences of arbitrary language and ad-hoc diagnosing. The history of psychiatry and the DSM is plagued with this pattern of inconsistency and drama, leaving questions about what can be truly known if the information is constantly changing—and how people can feel confident and safe with such unsteady fits and turns in the medical industry.

The DSM thus could be seen not as a final scientific document but rather as an ethnographic study—an account of how human beings are responding to the state and conditions of the world. In some eras, conditions related to slavery and homosexuality have featured more prominently; in others, depression and anxiety have dominated. As the DSM evolves, we are likely to see more "conditions" or "disorders" related to what is happening in the natural world, given the changes in land and ecosystems due to climate change and global warming. "Ecopsychology" is already a budding field.

If psychological framing is situational and contextual, what's certain is that the medical and psychiatric treatment of those who operate differently from a perceived norm tells a story of discrimination and pathologization. Greenberg asks, "If the people who write the DSM don't know which forms of suffering belong in it, and can't say why, then on what grounds can the next instance in which prejudice and oppression are cloaked in the doctor's white coat be recognized?"

History, language, context, and power are deep determinants of who gets framed as "normal" or "wrong." Sensitivity and sensitive women in particular have been no exception. The history of language within medicine and science has corrupted our notion and felt sense of what it means to be sensitive and thereby pathologized sensitivity and created an epidemic of shame among some of humanity's most gifted individuals. Let the reeducation begin.

## Chapter 2

# Reframing Sensitivity

A highly sensitive person with clear giftedness as well, Sherry holds multiple degrees in French, autism studies, and employment relations. "I didn't know about HSP until 2010 when I read Elaine Aron's book The Highly Sensitive Person," Sherry tells me over the phone one day. "I know I don't think like other people. I always make points that other people haven't considered or don't understand. Whereas a lot of people are just superficial in their decisions and judgments, I like to look at all the factors and consider all the angles before I make a determination."

She goes on to describe how bright lights give her headaches, which she's suffered since childhood, and how unexpected noises strongly affect her. Likewise for witnessing the strong emotions of others, which she says is very draining. For instance, she doesn't like

*Facebook, for fear that she will become too absorbed in the lives of her friends. "I need to stay within my limits," she tells me. And she limits the number of friends she has as a strategy to prevent overstimulation.*

*When she got her master's degree in French, she found that teaching students was also too draining. When her daughter was diagnosed with Asperger's, she decided to study autism as well. "I feel like I can relate to her and understand when she's overstimulated and what triggers her to have a meltdown, because I also have some of those difficulties." Sherry lives on a quiet street and practices yoga regularly; when things get to be too much, she goes into a dark room and closes the door. "There is an evolutionary advantage to these traits, and people need to value them," she tells me. "If you're like everyone else, then you're not going to be as innovative."*

## Elaine Aron, Queen of HSP

People described Elaine Aron as a shy, quiet, introverted child who feared putting her face underwater while swimming. She decided to study psychology partly in response to the dissonance she felt between her internal experience of the world and the way the world seemed to respond to her. Her studies and her interactions with her own clients helped her arrive at a framework for thinking about her particular makeup—she was highly sensitive.

The defining characteristic of HSP is a depth of processing—taking time to perceive and process external and internal input, be it sound, light, feelings, or new information and explanations. Aron found in her research that about 20 percent of the population is HSP to some degree (that number is split evenly between men and women, and 30 percent of HSPs are extroverts). Her coining of the term *highly sensitive person* has helped thousands of other therapists, researchers, and laypeople to better understand their own makeups and the makeups of those in their lives.

I don't remember the exact moment in college when I first found out about HSP, but as long as I've known the term, I knew I was highly sensitive. Over time, I followed Aron's work, and we later became friends and colleagues. In my interviews with her, Aron talked about how important it has been to keep the HSP idea and inquiry within the realm of science and academic research. Over the years other writers have put forth similar-sounding names and ideas, but Aron has been able to straddle the medical study of HSPs via brain imaging and maintain a clear narrative description of what the experience is like for people who identify as HSP.

"Having a sensitive nervous system is normal," Aron writes in *The Highly Sensitive Person*, "[it's] a basically neutral trait." In the book she details what the trait is, how to figure out whether you possess it, and ways to handle heightened sensitivity and overstimulation in a world that is often too fast, too bright, too loud, and simply too much. "What seems ordinary to others, like loud music or crowds, can be highly stimulating and thus stressful for HSPs."

Aron acknowledges the challenge of being sensitive and some of the stigma that may be attached to the label. But in her research she has found that when HSPs try to mimic others in what they perceive as "normal" or "expected" of them—such as tolerating bright lights and loud music—they become exhausted, burned out, depressed, or develop symptoms such as headaches and fatigue. This is because "what is moderately arousing for most people is highly arousing for HSPs." HSPs tend to shut down much faster than other people. But as in the case of many other neurodivergences, the gifts of HSPs are vast. They tend to excel in psychology, writing, art, and music and as entrepreneurs. Because their nervous systems are more attuned to subtleties in the environment, they excel in perception, detecting nuances, and understanding others.

In some of her recent studies, Aron has found that being more responsive to their environments may have served as a survival mechanism for HSPs historically and today. Brain imaging studies show that HSPs process stimuli more elaborately and that some brain regions are more active in the integration of sensory information, awareness, and empathy. Other studies by Aron and colleagues indicate that HSPs have a longer "pause to check" time, that is, they respond to new information by taking time to scan for nuanced details and thus take longer to reach conclusions or process information.

Statements from Aron's HSP questionnaire indicate some of the defining characteristics of the HSP:

- I am easily overwhelmed by strong sensory input.
- I seem to be aware of subtleties in my environment.
- Other people's moods affect me.
- I find myself needing to withdraw during busy days, into bed or into a darkened room or any place where I can have some privacy and relief from stimulation.
- I am particularly sensitive to the effects of caffeine.
- I am easily overwhelmed by things like bright lights, strong smells, coarse fabrics, or sirens close by.
- I have a rich, complex inner life.
- I am made uncomfortable by loud noises.

- I am deeply moved by the arts or music.
- My nervous system sometimes feels so frazzled that I just have to go off by myself.
- I am conscientious.
- I startle easily.
- I get rattled when I have a lot to do in a short amount of time.
- When people are uncomfortable in a physical environment, I tend to know what needs to be done to make it more comfortable (like changing the lighting or the seating).
- I am annoyed when people try to get me to do too many things at once.
- I try hard to avoid making mistakes or forgetting things.
- I make a point to avoid violent movies and TV shows.
- I become unpleasantly aroused when a lot is going on around me.
- Being very hungry creates a strong reaction in me, disrupting my concentration or mood.
- Changes in my life shake me up.
- I notice and enjoy delicate or fine scents, tastes, sounds, works of art.
- I find it unpleasant to have a lot going on at once.
- I make it a high priority to arrange my life to avoid upsetting or overwhelming situations.
- I am bothered by intense stimuli, like loud noises or chaotic scenes.
- When I must compete or be observed while performing a task, I become so nervous or shaky that I do much worse than I would otherwise.

Aron's research and coining of the term *HSP* was the first time that the quality of sensitivity was addressed in scientific literature without being pathologized, and it initiated a rapid-fire connection among women around the world who immediately saw themselves in her descriptions and consequently formed a kind of tribal bond of belonging under the banner of HSP. As Aron is a woman herself and her studies include many female participants, there was no risk of male-dominant research. The same goes for the language and vocabulary Aron uses and that is now widely accepted—feelings and experiences of sensitivity, as you can see from the list above, are stated neutrally without judgment.

## A Gaping Hole

Women have been absent from the growth of psychology as a field for the majority of its existence. That means that how we see the mind, talk about it, frame it, and ponder how to support it has been informed by thinking that originated in the minds of men, and by research that has largely been based on male subjects. Therefore, the theories, treatments, and general psychological approaches that have dominated Western psychology are male dominated.

This has become a particular concern for many traits, particularly autism and ADHD. The diagnosis rates among men surpass those of women, and even in studies on adolescents and children, girls are often left out of the research and discussions. Entire demographics of women remain hidden and invisible, even to their own selves. According to the World Health Organization, 450 million individuals have a diagnosis of "mental illness"—more than 6 percent of the global population—and that number is likely to be much higher given that mental health is a relatively neglected area of concern in many parts of the world and mental health symptoms of women often go completely unrecognized. This is a failure of modern scientific inquiry.

"Scientists aren't interested in differences because they want to arrive at theories," Aron tells me, "which means they want sameness. . . . Individual differences are looked down on in psychology because then you don't get your generalizable theories." And, she points out, the same goes for doctors and therapists, "because insurance companies dictate diagnoses and reimbursement." In other words, everyone needs to fit neatly into a category.

The very real problem, though, is that entire theories and diagnoses therefore are based on nongeneralizable sample populations because women have been left out of the picture. And because they have been left out of the picture, says Pauline Maki, director of Women's Mental Health Research and associate director of the Center for Research on Women and Gender at the University of Chicago, "oftentimes what women see as normal in the range of emotional experience has been pathologized."

It's no wonder that we women walk around with unnecessary amounts of shame, guilt, depression, and anxiety. Our reality has not been properly validated. This has important implications not only for sensitive neurodivergent women, but for the world as a whole. It's as though a human body (in this case society) has been operating without an arm, completely unaware of this other limb that has sensations, serves a purpose, and can exponentially improve the body's functioning. Sensitivity is a core human trait, fundamental to reciprocal relationships and building economies and growing society.

Of course, the problem of female dismissal and gender bias

is not limited to the field of psychology. Women have been routinely left out of studies within medical research more broadly. In her groundbreaking book *Doing Harm: The Truth About How Bad Medicine and Lazy Science Leave Women Dismissed, Misdiagnosed, and Sick*, Maya Dusenbery writes: "While their numbers have increased since the early nineties, when federal law began requiring women and racial minorities to be included in research funded by the National Institutes of Health (NIH), we are still feeling the legacy of years of foundational research conducted on men, with the assumption that it could be extrapolated to women." Dusenbery exposed this crisis in the context of people with autoimmune conditions, two-thirds of whom are women. Little research on women is being funded, and doctors are often clueless as to how to treat women with, for instance, chronic fatigue or fibromyalgia.

In her book *Inferior: How Science Got Women Wrong—and the New Research That's Rewriting the Story*, Angela Saini writes, "Until around 1990, it was common for medical trials to be carried out almost exclusively on men." This is partly because sticking with one sex in research samples is cheaper.

Multiple researchers, psychologists, and scientists told me this over and over—that women are systematically left out of research studies in favor of obtaining streamlined, publishable results that otherwise risk being tainted by the presence of women and their hormones.

## Shutting Sensitivity Down

"I can certainly speak from the perspective of being a woman inside of the system," the author Lissa Rankin tells me one day.

"The challenge of being an ob-gyn, a doctor of *women*—at the time I was going through medical training—is that it actually shamed, belittled, and rejected so-called feminine healing properties." In the beginning of her book *The Anatomy of a Calling*, Rankin describes how when she had to deliver four dead babies in one night, a male doctor literally chased her down the hallway screaming, "Buck up, Rankin! You'll never amount to anything in this business unless you can stop feeling so damn much!" From her perspective, her feelings of sadness and grief were a normal, healthy response to the situation. And the female midwives were the ones who were able to hold her and rock her when she fell to the ground sobbing with sadness for the grieving mothers and their babies.

"That felt really natural for me to be in an empathic response with my patients," she tells me, "and yet everything in the system teaches you to shut that down. We're taught from the beginning to not feel, don't take it on, don't talk about yourself, don't let anybody see how their experience is affecting you, be detached. And that way you can make rational, competent decisions and do your job."

What Rankin describes is characteristic of how the traits and qualities associated with sensitivity—and furthermore often associated with women more broadly throughout history—are neglected, covered up, and explicitly *unwelcome*. "That, to me, was representative of everything that was wrong with medical school," Rankin says. "My response was very normal—it was perfectly appropriate to have a fifteen-minute break so that I could feel what I was feeling and let it move through me. My body couldn't handle stuffing all that trauma."

Of note here is that this quality of sensitivity is found across so many different types of people that one wonders why the label is even necessary. The answer to that is because of the way sensitivity is not honored as part of our human experience, and hence the gift that it offers is not let loose into the world. But as science catches up to what so many women have known for centuries—that we often sense feelings and emotions—people can more fully claim their gifts, stop hiding, and be more visible across academia, medicine, education, psychology, psychiatry, and the sciences in general.

"It would have been so different had any of my teachers simply said, 'That was tough what you just went through and you're having strong emotions. Me, too,'" Rankin says. "That's what good doctors do—they feel things and they're intimate and it's a real privilege. This is what made me want to be a doctor and a healer."

## Toward a More Sensitive Medical System

Rankin ultimately left mainstream medicine and launched Whole Health Medicine Institute, in addition to authoring several bestselling books. Her initiative and approach are good examples of how we might better integrate the concerns, needs, and wishes of women generally, and sensitive women in particular. Rankin says: "In my Whole Health Medicine Institute we ask women, 'How did you violate yourself in order to become a doctor?' And the stories they tell are of deep trauma—unbelievable trauma. I had several trauma therapists present who had helped war veterans and trafficking victims—and they said, '*We've never seen this much trauma.*'"

When I heard this, I wasn't surprised. There is a difference between traumatic events and trauma. Many think of war or rape when they hear those words, and rightly so, but there is another form of "silent" trauma that has to do with the violation of an entire demographic of people via cultural structures. If the perspective of women has been missing in action from setting agendas, determining work cultures, making decisions, and setting the tone for how we operate—that is most certainly a form of trauma. It is violation and oppression, and neuro-divergent women—particularly those who are sensitive—suffer the acute consequences because of their heightened awareness and processing. And it's even harder for women who are also marginalized because of their race and class.

Author angel Kyodo williams is one such woman. Growing up as a sensitive, gifted, quiet, book-oriented young black woman, she had to learn how to manage the profound empathy she felt toward others in her New York City neighborhood and housing development. The expression of sensitivity is a challenge in Western culture to begin with, and for many marginalized communities sensitivity may be viewed as even more of a challenge. Later as an adult, williams was also one of the first women of color to be included in a study on the autoimmune condition lupus. It has taken decades to integrate the concerns of women, let alone women of color, into scientific and medical research. There is no end to these stories of bright, sensitive women whose needs are foreign to the medical and research establishments.

"Our entire culture is so sick," Rankin tells me, "whether you're talking about the media, legal system, politics, education, or health." Rankin underscores how out of balance we've become about suppressing and minimizing sensitivity. But she also acknowledges that how we operate and function and interact are not set in stone; human beings are teachable, and there are other ways of living. In her case, she has found tremendous insight and practical wisdom from observing how people live in other countries and cultures, and that has helped to restore balance in her own life. She has the humility to see that the evolution of the Western medical system is not the singular evolution of all approaches to health and well-being; for example, she has studied and incorporated South American and Asian frameworks into how she relates to her clients.

As Rankin and I talked, I gradually shared more about my personal story. I had only barely mentioned ADHD and was discussing women and mental health more broadly. But then out of nowhere she told me, "I'm a magnet for people on the autism spectrum—my roommate, ex-boyfriend, and teacher are all autistic." "Wow," I thought, "this is so interesting, and at the same time not surprising." I had struck up a conversation with her based on what I knew of her combination of the hard sciences and more alternative viewpoints, which intrigued me. Perhaps it makes sense that someone so comfortable at the edge of defined paradigms would also naturally have folks in her circle who identify as neurodivergent.

## Sensitivity, Hysteria, and What It All Means for Women

As we have seen, "madness" and "hysteria" occupied the attention of doctors and scientists from the 1600s onward, and the phenomena were mostly observed in women. Sensitivity—or "emotional reactivity," as it has been dubbed historically—is core to hysteria, so we must not ignore how history has pathologized both sensitivity and women. To put it plainly, if women have been painted as sensitive, and sensitive is bad, then does that mean women are bad?

Likewise, if men dominated medicine and science—particularly psychology and psychiatry—for hundreds of years and shaped its foundations and outcomes, then it's logical to conclude that a female imprint is lacking. That is to say, we women have not made our mark, the disciplines don't reflect our realities, and it's possible that our very existence is pathologized. Had our realities been welcomed and embraced throughout the evolution of science and medicine, the narrative around sensitivity would look very different today.

Sensitivity today is well understood within particular disciplines but simultaneously is not well understood as a whole. The majority of research available regarding sensitivity is limited to individual studies, as few books have been written on the subject. In the 1970s, occupational therapist A. Jean Ayres came up with sensory integration theory in response to rising rates of sensory processing challenges. Neurophysiologist Giacomo Rizzolatti first labeled what he called "mirror neurons" in 1980, which gave us an idea about the neurons responsible for mimicry and empathy. As we have seen, Elaine Aron coined the term HSP in 1996 as a way to describe a more extensive depth of processing that occurs for about 20 percent of the population. Most recently, researchers at the University of California, San Francisco, have been studying sensory processing disorder through the SPD consortium registry and database.

But within most of the literature that exists about sensitivity, the words *disorder* and *abnormal* are used extensively. There seems to be a constant comparison to a supposed "norm" regarding how people perceive and process sensory input. *Sensory processing* is the common technical term used in studies, occupational therapy guides, and elsewhere to describe how sensitivity is measured—how individuals process stimuli. Within focused areas of the study of sensitivity, multiple subtypes of sensitivity are often described in detail, which furthers the case that a spectrum of sensitivity is normal and expected within any given population and the words *disorder* and *abnormal* are thus unnecessary.

One begins to wonder how sensitivity has become pathologized over time. Furthermore, the prevalence of sensitivity across so many neurodivergences encourages us to consider what its pathologization reflects about our world. Do we live in unnatural ways? Is everyday life set up to cater to only those who are not sensitive? And if so, what does it mean that the structures, leaders, and rhythms of our lives are dictated and shaped by influences that don't honor sensitivity?

## Sensitivity as Opportunity

Surprisingly, the majority of research we have on sensitivity is from studies on animals. But much of what you might read on sensitivity among animals is fascinating and will likely resonate with you in human terms.

Howard C. Hughes was fascinated with animals and unseen sensory systems, so he wrote *Sensory Exotica: A World Beyond Human Experience* (1999), diving deep into the unseen lives of animals. Like other scientists who have looked to monkeys and horses, for instance, to better understand animal communication through sensing, Hughes turns to animals to help infer what might be happening in humans and what might be *possible* for humans. Just as occupational therapists refer to interoception as an internal sense of body awareness, Hughes explores other potential senses. This leads to the question: For neurodivergent people, what extra information are they perceiving or sensing that neurotypical people are not?

As examples, Hughes points to animals that have internal compasses that inform their navigation across long distances. Whales communicate with other whales across miles of open water, and tiny bats possess sonar systems far superior to those of advanced human-created submarines. For humans, "it is difficult to appreciate the extensive computations that underlie even the simplest sensory experience," writes Hughes. Our vision takes in only a tiny fraction of the full electromagnetic spectrum. And some animals, he notes, "are endowed with different types of receptors that render them sensitive to portions of the spectrum that we cannot see." Whales and bats have two operating modes of hearing—a passive mode to detect external sounds, and an active, biosonar mode that "relies on reflections of self-produced sonar signals." Would we consider this a disorder? Surely these sensory abilities are incredible strengths—or at the very least basic survival skills. So why do we think of different sensory abilities as pathological when present in humans?

Hughes notes that there were misperceptions and misinterpretations about animal sensory systems along the way that delayed understanding for hundreds of years. "Sometimes it is our own parochialism, our own narrow frame of reference, that provides the biggest obstacle," he writes.

We don't want that same delay for our human species. Right now, we have the opportunity to remedy egregious mistakes and oversights in how we understand, frame, describe, and respond to human difference. Robert Whitaker, the author of the book *Mad in America*, has launched a website that includes blogs of personal stories about the perils of modern psychiatry. Named Mad in America: Science, Psychiatry and Social Justice, the website allows people to share and reframe their experiences of mental differences and seeks for the wider public to do the same.

One story that I came across on Mad in America seemed to understand this idea that sensitivity is something to be harnessed and cherished, and that the trait may be the *antidote* to society's modern ills. The author describes "sensitives" as potential problem-solvers and urges that we as a society keep sensitives healthy and intact. This is essential to flipping the script on sensitivity and neurodivergence—that is, thinking about the modifications we can make to how our world operates *rather than to the individuals who make up our world.* And, since women have been so fundamentally missing from helping to shape larger societal structures and the fields of psychology and medicine, it's likely that inclusion holds the key to helping to bring wholeness to a fractured culture, a softening of our defenses against sensitivity, and a thriving life to neurodivergent women—and men—everywhere.

# Part II

# OUTER FRAMES

## Chapter 3

# Autism, Synesthesia, and ADHD

C Hart grew up in California's Central Valley with a mother who noticed her strong body awareness as a child. Hart's mother, a nurse, encouraged her daughter to go into physical therapy in order to use her hands, and eventually Hart became a massage therapist. A client once mentioned synesthesia to her, and a kind of relief and shock hit her all at once. She finally had a word to describe her experience of feeling what others feel, seeing colors when reading books, and more. She didn't learn she had synesthesia until she was in her early forties.

Hart was a gifted child who began reading at age two and

*a half, and she had always seen letters as colors and symbols. A "mirror touch" synesthete as well, she would feel others' pain, both physical and emotional. "I would see another child's scrape, and I would feel a shock of electricity go up and down my body. It never occurred to me to say anything because this was just my experience of 'normal.' I didn't know that other people didn't experience that."*

*Hart's sensory sensitivities are varied, and she wonders whether she might be classified on the autism spectrum as well. She was recently diagnosed with ADHD, something she had suspected and that was suggested to her after finally finding a psychiatrist who took the necessary time to analyze and interpret the nuances of her "symptom presentation" beyond stereotypical ideas. For many people, autism, ADHD, and synesthesia occur together. She describes her world as "too much or too loud."*

*Now as an adult, Hart walks into a room and is a magnet for others, with her long red hair and flowing colorful outfits. When I met her one evening and she hugged me close, I felt that same shock of electricity she describes. Her warmth is palpable. But when Hart was an infant, her mother took her to a pediatrician because she thought something was wrong when she didn't engage with others like her sister did. These are themes that many women shared with me over the course of my writing this book—being bright, having multiple sensitivities or diagnoses, and having parents who fear that something is very wrong. For those of us like Hart, it takes time for our nervous systems to become calibrated to the "outside" world and the staccato rhythms that feel like assaults. After a series of trials and errors, figuring out what type of work matches our skills and needs or what kind of people are compatible with us, we finally come out of our shells.*

## Autism

A person with so-called classic autism is what many doctors and laypeople describe as "socially awkward," "in their own world," and lacking empathy and "normal" or "appropriate" social interaction. This pathologizing language describes a person in terms of individual norms and expectations; we don't often stop to think about what is happening in society on a larger scale that would generate such a description or make such a person sound so "bad." Only when we step back and examine the behaviors from a nonstigmatized perspective can we realize that people described this way aren't *bad*; they are merely *different* when measured with an established (and some might say arbitrary) barometer. In fact, many people in the autistic community like to flip the script and point out how awful neurotypical culture and expectations are: think small talk, social niceties, herd mentality, compliance, and other unpleasant or taxing behaviors that are deemed "normal."

While figures like the professor and author Temple Grandin have brought attention to the gifts of autism, our culture as a whole remains steeped in images and definitions that see this kind of neurological makeup as a defect. More and more we are appreciating the fact that the range of autistic experience is vast, necessitating the use of the label *spectrum*; and autistic advocates are now speaking out and challenging pervasive false notions.

This is most apparent on Twitter, where autistic culture thrives and autistic individuals all over the world share personal stories and insights on their own terms and in their own voices. "Nothing about us without us" is a popular slogan. Instead of being studied from the outside, autistic people are giving voice to their experiences on their own terms. Twitter is a perfect vehicle for this, where no stone is left unturned and no question is too cumbersome: Why is the rate of suicidal ideation higher among the autistic population? Why do autistic individuals prefer that the month of April be called "autism acceptance" month rather than "autism awareness" month? The idea that autistic people lack empathy, for example, is often countered on Twitter, with people sharing how in fact it's their *overabundance* of empathy that causes them to shut down emotionally and retreat. There's no lack of empathy, but too much of it. How's that for reframing?

What I heard in researching, conducting interviews, and reporting on the autistic experience is that sensitivity features prominently in autism and is both a strength and a challenge. In the case of autistic women, many of whom have been sensitive their entire lives but not known they were on the autism spectrum, the question of utilizing and managing sensitivity is tricky because many have been told or taught that something is wrong with them. As a result, many have experienced shame, depression, and severe anxiety.

Author Samantha Craft, who is autistic and is often a go-to resource for women wondering whether they are also autistic, has compiled an unofficial list that she says is "a springboard for discussion and more awareness into the female experience of autism." This can be helpful because the authors of the DSM are often changing its scope, and many doctors and therapists are at a loss regarding how to diagnose. Here are some abbreviated items from her list regarding autism/ Asperger's:

| | |
|---|---|
| Seeing things at multiple levels, including her own thinking processes | Escaping in thought or action to survive overwhelming emotions and senses |
| Continually analyzing existence, the meaning of life, and everything | Escaping through fixations, obsessions, and overinterest in subjects; through imagination, fantasy, daydreaming, and mental processing; through the rhythm of words |
| Often getting lost in her own thoughts and "checking out" (blank stare) | Imitating people on television or in movies |
| Experiencing trouble with lying | Escaping through relationships (imagined or real) |
| Finding it difficult to understand some human characteristics, such as manipulation, disloyalty, vindictive behavior, and retaliation | Escaping into other rooms at parties |
| Experiencing feelings of confusion and being overwhelmed | Being unable to relax or rest without having many thoughts |
| Experiencing feelings of being misplaced and/or from another planet | Feeling extreme relief when she doesn't have to go anywhere, talk to anyone, answer calls, or leave the house but at the same time harboring guilt for "hibernating" and not doing "what everyone else is doing" |

| | |
|---|---|
| Feeling isolated | Perceiving visitors at the home as a threat (this can even be a familiar family member); knowing logically that visitors are not a threat but still feeling anxious |
| Obsessing about the potentiality of a relationship with someone, particularly a love interest or new friendship | Dreading upcoming events and appointments |
| Being confused by the rules of accurate eye contact, tone of voice, proximity of body, body stance, and posture in conversation | Feeling anxiety knowing she has to leave the house and feeling overwhelmed and exhausted by the steps involved |
| Becoming exhausted by conversations | Preparing mentally for outings, excursions, meetings, and appointments, often days before a scheduled event |
| Continually questioning the actions and behaviors of herself and others | Continually questioning next steps and movements |
| Training herself in social interactions through reading and studying other people | Feeling as if she is on stage being watched and/or having a sense of always needing to act out the "right" steps |
| Visualizing and practicing how to act around others | Having huge compassion for suffering (sometimes for inanimate objects) |

| | |
|---|---|
| Practicing/rehearsing what to say to another person before entering a room | Being sensitive to substances (environmental toxins, foods, alcohol, medication, hormones, etc.) |
| Having difficulty filtering out background noise when talking to others | Questioning her purpose in life and how to be a "better" person |
| Having a continuous inner dialogue about what to say and how to act in social situations | Seeking to understand personal abilities, skills, and gifts |
| Feeling great peace of mind when she knows she can stay at home all day | Feeling trapped between wanting to be herself and wanting to fit in |
| Requiring a large amount of downtime or alone time | Imitating others without realizing it |
| Feeling guilty after spending a lot of time on a special interest | Rejecting and/or questioning social norms |
| Disliking being in a crowded mall, gym, or theater | Having a hard time feeling good about herself |
| Being sensitive to sounds, textures, temperature, and/or smells when trying to sleep | Not understanding jokes |
| Adjusting bedclothes, bedding, and/or environment in an attempt to find comfort | Remembering details about someone's life or details generally |
| Longing to be seen, heard, and understood | Relieving anxiety by writing or being creative |

| | |
|---|---|
| Questioning whether she is "normal" | Having certain "feelings" or emotions toward words and/or numbers |
| At times adapting her view of life or actions on the basis of others' opinions or words | Experiencing simple tasks as causing extreme hardship |
| Viewing many things as an extension of herself | Feeling emotionally challenged by new places |
| Disliking words and events that hurt animals and people | Feeling a sense of panic when faced with anything that requires a reasonable number of steps or amount of dexterity or know-how |
| Expecting that by acting a certain way certain results can be achieved, but realizing in dealing with emotions, that those results don't always manifest | Feeling anxious at the thought of repairing, fixing, or locating something |
| Believing that everything has a purpose | Avoiding mundane tasks |
| Having a difficult time making or keeping friends | Feeling overwhelmed by something as "simple" as a trip to the grocery store |
| Having a tendency to overshare | Having a hard time finding certain objects in the house but clearly knowing where other objects are; feeling anxious when unable to locate something or when thinking about locating something (object permanence challenges) |

| | |
|---|---|
| Lacking impulse control when speaking (at a younger age) | Perceiving situations and conversations as black or white |
| Monopolizing conversations and speaking exclusively about herself | Overlooking or misunderstanding the middle spectrum of outcomes, events, and emotions (having an all-or-nothing mentality) |
| Having difficulty recognizing how extreme emotions (outrage, deep love) will affect her and being challenged to transfer what she has learned about emotions from one situation to the next | Overreacting: a small fight might signal the end of a relationship or collapse of the world; a small compliment might lead to a state of bliss |
| Always trying to communicate "correctly" | Wanting to know word origins and/or origins of historical facts, root causes, and foundations |
| Having trouble identifying feelings unless they are extreme | Noticing patterns frequently |
| More easily identifying personal feelings of anger, outrage, deep love, fear, giddiness, and anticipation than emotions of joy, satisfaction, calmness, and serenity | Remembering things visually |
| Seeming to others to be narcissistic and controlling | |

Craft also lists frequent co-occurring attributes, including obsessive compulsive disorder (OCD), sensory issues (sight, sound, texture, smells, taste—synesthesia), generalized anxiety, feelings of polar extremes (depressed/overjoyed; inconsiderate/oversensitive), and chronic fatigue and/or immune challenges. And as we have seen, an autistic person might be (mis)diagnosed as having a "mental illness."

Autism is thus a living, breathing group of attributes that gets missed as a diagnosis because it is broad, overlaps with many other traits, and is steeped in media-driven stereotypes. Many women are thus self-diagnosed. For those who get a clinician's affirmation, many different diagnoses may have been considered, with autism as a kind of final question on the table. Therefore, many women don't have an exhaustive understanding of themselves until later in their lives.

## Reaching for the Stars

"I've never been officially diagnosed. It's on my list of things to do," Sara Seager tells me. A planetary scientist and astrophysicist at the Massachusetts Institute of Technology, Seager searches for exoplanets, but has trouble with grocery shopping. "Someone wrote an article about me strongly suggesting I had Asperger's, and then one of my mentors read it. His wife was one of the first doctors of autism, and he called me and told me his wife said that I was definitely on the autism spectrum. At first I said, 'No, I don't have that.' And then I thought about it, and everything came together.

"When I was a child, other kids always thought I was odd. One of my friends who's definitely Asperger's never interacted with children when he was little. I wasn't as extreme as him, but before I found out I had Asperger's, I noticed he always made me laugh. One time we were meeting for pizza and I was five minutes late, and he's like, 'What's the protocol if someone's late?' He honestly didn't know, and he had to have it all written down, like how many minutes to wait before it's rude to message them. So I watched him navigate the world, and it was really ironic that later I had it."

Seager has received a MacArthur Fellowship, a rare and prestigious prize awarded to a select few. She has been featured in books, talks, and academic panels, and at the time of our interview, she had just signed a seven-figure book deal. She tells me that many kids had crushes on her at school. That made her feel "normal," but other than that, looking back, she felt very different—until she arrived at MIT. Now she's in her element and describes a sense of belonging and fitting in.

"When I look back on my whole life, things just fit together. My sister told me my life was really rigid and scheduled," she laughs. "If the schedule deviates from anything, I get really uncomfortable and my day will almost fall apart." She shares with me the confusion and frustration that ensue when a schedule changes, someone doesn't stick to a plan, or other unforeseen alterations occur. I resonate with her experience as I listen.

Seager's days are structured and busy. She's often at work by 6:30 or 7:00 a.m. and gets a whole day's work done on her own research in two hours. Then she spends the rest of the day teaching classes or attending meetings. When she gets home in the afternoon, she walks the dog and makes calls to the West Coast. By 5:00 p.m. she drives her kids to sports activities.

But it took her several years to streamline grocery shopping, something she tells me she thinks she finally has a handle on.

She used to go to the store and ask, "Where are the apples?" Although she thought she was using her normal voice, others would think she was being rude and abrupt and yelling at them. She says that people—her "interface with the world," as she puts it—provide the most challenge. "I have to slow down and make small talk first," she says. One time her neighbor even took time to explain to her how people warm up, slowly get to the point, ask questions, and then gently move away. At the nearby dog park, Seager can get annoyed when a fellow dog lover takes ages to get to the point of asking for a signature on a petition. She wants interactions to be clear, direct, and efficient.

She shares the delight and curiosity that has come with figuring out things like the furnace in her house and "detecting" when it's about to break down by "listening to the sound." But she has a persistent sense that things have been made more complex than necessary. I have found myself wondering the same, unsure why simple daily tasks are not more clearly taught or explained.

Of her particular sensitivities, Seager says, "I don't like being hugged or touched, even now. It feels like more of an awkwardness. I never cast it in terms of a sensitivity, but maybe it is." When she was little, her father—who was a doctor—thought she was "mentally retarded," in her father's words, because she often stared off into space. "I also cried for the first two years of life," she tells me. "I didn't get a lot more information than that, and he's not alive now so I can't probe him, but he thought something was wrong with me."

Seager says that "at MIT it seems like most people are somewhere on the spectrum. I feel at home there. And many people [at MIT] say it's the first time they have had friends." People are okay with her straightforward style, which makes her life easier, as it takes away the barrier of having to cushion language with small talk. She can let her guard down at MIT, and many on her team are very "Asperger's-ey," as she puts it. People there tolerate her communication style well, but Seager finds dealing with the outside, neurotypical world to be exhausting. It takes a huge amount of her emotional energy to think through everything before she speaks or acts. "It's just tiring," she says.

Seager relays to me her realizations about pacing in conversation, "shooting the breeze," and other social formalities. Sometimes she waits patiently while others take time to understand something, but to her it feels like "watching moss grow." She'll often call her husband at work to see how he's doing, and then when she feels she's done she'll suddenly say, "Okay bye," and hang up. "My husband used to call me a narcissist," she tells me, "but when this whole Asperger's thing came about, he stopped. It helped us tremendously because I think it helps him realize that when I'm being dismissive of something, it's not because I'm rude. It's just how I am. And it helps me realize that he's very sensitive and I've got to be aware of it. Now that it's more of an open thing between us, it helps a great deal."

Ultimately, Seager says, "Being on the spectrum contributed to my success." She attributes much of that success to her not being bothered by the same superficial concerns as her peers when she was growing up. The life of her mind shielded her from the necessities of small talk and inefficient social norms that require many adolescents to sink time and energy into "fitting in." Now as an adult she has two children who deeply understand her and a loving relationship with her husband. And she loves her work. "I always loved the stars and astronomy and thinking about what's out there," Seager says. "I was good at math and physics and was able to put it all together. I was lucky I was good at something I loved."

## Synesthesia

Synesthesia is a well-documented phenomenon of the brain whereby an individual's senses get "crossed" such that hearing sounds may elicit a visual field of colors, for example. A particular strand of synesthesia is called "mirror touch synesthesia," that is, a person can feel what another person feels by simply observing what is happening to the other person.

The NPR podcast *Invisibilia* documented one woman's experience with synesthesia and described in detail how she had to keep herself isolated because of the severity of her trait. One time while in a grocery store she fell helplessly to the ground when a little boy fell nearby. Other times she becomes overwhelmed by the emotions of her children or others in close proximity, which some therapists refer to as "mirror emotion synesthesia."

Synesthesia is the topic of many studies. A 2011 study in the *Journal of Neuroscience* reported that synesthetes were superior in facial expression recognition but not facial identity recognition. The researchers concluded that mirror touch synesthetes may possess enhanced emotion processing and call to attention the somatosensory system in the process of emotional empathy and perception of expression. A 2007 study in *Nature Neuroscience* found that participants reported a higher percentage of mirror touch errors as compared with control subjects—that is, synesthetes more often mistook perceived touch for actual touch because they feel the same to them. And a 2013 study in the journal *Emotion* reported that mirror touch synesthetes had a heightened and accurate recognition response to the emotion of fear, but not to those of happiness or disgust. Like Elaine Aron's characterization of HSP as an evolutionary survival strategy, the researchers concluded that relying on somatosensory mechanisms would be evolutionarily adaptive—explaining the existence and persistence of mirror touch synesthesia in humans.

CC Hart, the synesthete we met earlier, told me that when she was a kid, she could "tell you what page a certain scene was on because all the pages had corresponding colors." This was an experience she had known her whole life but lacked a name for. Finding out there was a name for her experience and that others in the world shared it gave her a feeling of community and connection. "So much opened up, and it was a huge sense of relief in recognizing that there were communities of people like me. The sense that others have had the same experiences helped me recalibrate what it means to be a neurodivergent individual." She is now a board member of the International Association of Synaesthetes, Artists, and Scientists (IASAS).

## Digging Deeper, Parsing Out

Many people relate to the experience of sensing others, emotions, and the environment around them—in fact many of us simply take this to be a natural part of who we are. Now studies on mirror neurons, empathy, and neurodivergence support the experience and give it a much-needed lens inside of academia. This sensitivity is also shown in animal species such as monkeys, birds, rodents, and fish; studies describe select members of a species displaying increased "responsivity," "flexibility," and "plasticity," indicating an evolutionary advantage. In more recent years this field has come to be known as "sensory biology"—with dedicated researchers at Johns Hopkins, Duke, the University of California campuses, and elsewhere—and has maintained a focus on animals.

But what does this all mean in human terms? That's where mirror neurons come in.

In his book *Mirroring People: The New Science of How We Connect with Others*, Marco Iacoboni writes: "The human brain contains about one hundred billion neurons, each of which can make contact with thousands, even tens of thousands, of other neurons. These contacts, or synapses, are the means by which neurons communicate with one another, and their number is staggering." Iacoboni was a student of Giacomo Rizzolatti, the Ukrainian scientist who first "discovered" mirror neurons while studying monkeys in his laboratory in Italy. Rizzolatti's lab focused on the neocortex, specifically a region of the neocortex labeled F5 that is responsible for planning, selecting, and executing actions. The story goes that as a scientist reached for something, a burst of activity was recorded coming from an observing monkey's brain, even though the monkey didn't move or copy the scientist's gesture. This happened several times. "Neither they nor any neuroscientists in the world could have imagined that motor cells could fire merely at the *perception* of somebody else's actions, with no motor action involved at all," Iacoboni writes. "Cells in the monkey brain that send signals to other cells that are anatomically connected to muscles have no business firing when the monkey is completely still, hands in lap, watching somebody else's actions. And yet they did."

At the time, neuroscientists believed that specific functions controlled by the brain were limited to specific "boxes." In line with earlier mechanistic thinking that has shaped the study of the brain, such a belief makes sense. As Iacoboni explains it, neuroscientists have tended to think of perception, cognition, and action as distinct spheres. But to begin to entertain the possibility that there is crossover and more complexity is exciting. Further experiments found that close proximity also determined a mirror response and that simulations would not elicit it—the objects or people executing an action had to be *real*, not robots or graphics, in order to elicit a mirror response. The tactile receptive field and the visual receptive field were found to be related in the brain. A new idea was thus born within neuroscience whereby some researchers concluded that the brain was attempting to create a map that took into consideration the visual and tactile space surrounding the body and that accounted for potential actions within that space.

This discovery and the evolution of our understanding of mirror neurons is important to situating a neurodiversity framework within modern life. The research scientists in Italy were intrigued and perplexed and looked to earlier philosophical explanations of human behavior, action, empathy, and learning. Rizzolatti and his colleagues departed from much of the history of the science of the brain, venturing out and becoming known for their range of insight and inquiry. This way of thinking could prime the psychological sciences for how to think about neurodiversity, given its natural focus on variety, as opposed to a mechanistic or reductionist, disease-based model of mental differences.

## The Synthesizing Synesthete

Harvard neurologist Joel Salinas has synesthesia and documents the overlap with other traits such as autism in his 2017 book *Mirror Touch: A Memoir of Synesthesia and the Secret Life of the Brain*. Salinas confirmed to me the unique interplay between genes, biology, and experience that gives rise to such overlapping sensitive traits or diagnoses. "People on the autism spectrum are much more likely to have synesthesia compared to the general population," he says. Studies confirm this as well.

"Could you call somebody with synesthesia autistic?" he asks rhetorically as we speak. "That ends up being more a matter of semantics and technicalities, because these are all concepts. What we're trying to describe when we talk about synesthesia is a sensory experience. So someone who has synesthesia may also fall in the definition of the autism spectrum, but someone with synesthesia may not necessarily meet the checkboxes or criteria to be defined as someone on the autism spectrum."

I find Salinas's technical prowess reassuring while talking to him, as his nuanced understandings and insights surpass those of most psychologists and psychiatrists. A "behavioral neurologist," he studies the dynamic relationship between social relationships, genes, and neurological traits. "Many of the genes that have been identified to be linked to synesthesia have overlap in the genes that have been identified as strong candidates or likely involved in the autism spectrum," he tells me. Differences in brain connectivity are found in both autistic people and synesthetes, explaining why both groups experience similar sensations. An additional finding is that the way autistic people *describe* their experiences of sensation and sensory differences overlaps with the descriptions of people who have synesthesia.

For instance, CC Hart says, "I wear headphones on the train, the same ones that DJs wear to block out sound. I have trouble filtering out ambient sound from the sound I might be focusing on, like the people sitting next to me. Sounds also have colors or feelings in my body, and I feel like I'm getting battered when I'm in a really sensory-rich environment." This way of describing her sensory environment is almost identical to the way many autistic women describe their lives. "Learning about synesthesia was a gateway to recognizing the issues I've always had with processing sensory feedback, especially tactile and auditory," Hart says. "I've got a low level of olfactory tolerance when things smell really strong. My tolerance for crowds is low, my tolerance for aromas is low, and my tolerance for sound is low and I take it in bits and pieces."

Salinas's perspective as a medical professional who is also a synesthete provides fascinating insight as well. I reached out to him after reading his book, and he flew to San Francisco from Boston to do an event with The Neurodiversity Project, the community forum I host for authors exploring innovative research and solutions in medicine and society. An extremely articulate and eloquent individual, he quickly earned my respect for both his warm manner and sharp scientific insight.

Growing up in Miami, with a short stint in Nicaragua, Salinas experienced sensations intensely and never had a name for the colors and empathy that filled his daily life. While in India on a medical school trip, someone mentioned synesthesia, and finally everything clicked for him. In his book he details what it's like to perform surgery or give a hug or treat a psychiatric patient as a synesthete—suffice to say he feels everything intensely so he has had to both embrace and manage the experience. Feeling profound empathy and feeling overwhelmed are both part of the picture for him, and he has expertly learned to set appropriate boundaries for himself and others. In my opinion, part of what makes him such a popular neurologist at Harvard and Massachusetts General Hospital is that he embodies the neurodiversity framework as he relates to his fellow neurodivergent patients. He

understands and embraces difference, feels no need to pathologize, and seeks to "treat" people only when they express distress
and a desire for treatment.

"Is synesthesia a gift, a curse, or none of the above?" Salinas
asked at a recent conference, and he tells me that the same could
be asked not only of the autism spectrum, synesthesia, or other
traits, but of brains in general. "Is having your brain a gift, a
curse, or none of the above?" he reiterates. Context is key, he
says, because in one situation, one set of traits can be highly
beneficial, and in another situation, those same traits can be
"crippling."

As an example, he tells me about a woman who has obsessive
compulsive disorder as defined in the DSM. She is highly obsessive and can barely function in situations where she has to make
quick decisions. But she is a genius in the operating room, where
she is a scrub tech—so she is the one organizing the instruments
for surgeons at exactly the right time. She is very meticulous,
partly because of how her brain is wired and programmed.
So what would be an impediment in some situations is an advantage in the operating room. "So it really comes down to
context," he says, "which is why the DSM and those definitions need to be contextualized. The function of the DSM is to
give a name when someone is not doing well, meaning they're
having trouble functioning with their occupation or socially—
basically they're in distress in one way or another. And it's the
*patient* relating that they are in distress, not a family member
who is in distress over the other person's behavior."

The clinician's responsibility is to dig down deep and determine whether the environment is the problem or whether something is happening with the brain and body. "That's the goal
of medicine—or should be," Salinas says. "My sense is not to

use the DSM as a means to label people, because that can lead to all sorts of alienation and loneliness, and that in turn can create problems; but to use the DSM as a way of helping the person in distress. Some people benefit from having a label, but not everybody does." He tells me about one patient who had a number of cognitive symptoms of depression, and when he explained that to her, she began to cry because she was so grateful to have a name for her experience. "It's in those situations that the DSM can be useful, but when the DSM is used to stigmatize or [its categories are] imposed upon somebody, I think that's when it can be problematic."

Salinas thinks the DSM is a broad and primitive tool used to understand the brain and believes a much better understanding of the brain will come from focusing on the interplay between biology and the environment. "I think psychiatry is moving there," he says, "which means it's also moving closer toward neurology." People are moving away from checkboxes of observations and instead focusing on patterns of physiology. For example, the brain tends to react this way or that way in such-and-such a context. Or your brain appears $X$ or $Y$ way, or your brain connectivity looks like $A$ or $B$, or you have this collection of genes. "I think that will lead to better insights about what the underlying cause is, and the more specific you get about what the underlying cause is, then the more specific you can get about treatments for the people who *are* in distress."

Since Salinas is a behavioral neurologist, one of his goals is to figure out what his clients are communicating. "We're all living in these bodies that we're born with, outside of our own choice, and our bodies come with all sorts of weird things, and our brain is locked in this dark tomb and there's no way to really know what another person is experiencing." Language is our primary communication tool, which of course has limitations. "So trying to tease apart in as granular a way as possible what someone is trying to communicate—I find that extremely helpful."

## The Third Factor: ADHD

The co-occurrence of autism and synesthesia is well documented, but the sensitivities found within autism and ADHD are even more well known. For autistics, sensory sensitivities often dictate the development of particular behaviors. *Stimming* refers to movements—such as flapping the hands or tapping the fingers—that help relieve anxiety that comes with overstimulation. It can also take the form of mental stimming, such as repeating numbers, words, or letters (also referred to as *echolalia*). When I was a child, for instance, I had a mental count of the electrical poles on my street, and every time we drove by, I had to mark in my mind the midpoint between each pole.

With ADHD the sensitivity shows up differently. High stimulation is both exciting and confusing for people with ADHD, because they can get overwhelmed and overstimulated easily without realizing they are approaching that point. Along with sensory regulation, emotional regulation becomes difficult, which accounts for some of the sensory overload, or "meltdowns," common in ADHD, as with the autist. There is a shared experience across both neurodivergences of gradual sensory overwhelm—and then having to recover.

A 2006 report from the National Autistic Society indicates that autistics have a lower threshold for audio, visual, oral, and touch stimuli. Such "abnormalities" were found to decrease over time, except for touch sensitivity. A 2014 study in the *American Journal of Psychiatry* documents the emotional "dysregulation" that occurs for 30–70 percent of people with ADHD, making emotional sensitivity a higher level of concern than generally considered for this population. Sensitivity is core to both of these neurodivergences and gets masked by the use of words such as *dysregulation*, *abnormality*, and *dysfunction*.

ADHD is commonly thought of as something that makes little boys fidgety, distracted, and unable to stay still. Doctors and researchers often give little attention to what is happening under the surface and in the broader environment. And, of course, there is very little focus on girls and women. As in the case of autism, our outdated stereotypes are based largely on an initial excitement and eagerness in the medical community to capture and frame a so-called new set of behaviors that started surfacing. And because so many girls and women with ADHD are "smart" and have done well in school, they've flown under the diagnostic and research radars. Women with ADHD who have struggled for years with logistical challenges often develop a nagging sense of not being good enough, never being able to "hit the mark" at work or home, and they struggle with anxiety and depression. But many women with ADHD also use their gift for hyperfocusing to excel beyond their peers—in writing, research, art, and other areas. Remember, despite what the words "attention deficit" imply, ADHD is not a *deficit* of attention, but rather a challenge of regulating it at will or on demand. People with ADHD often have *too much* attention—just not at the "socially acceptable" times or situations found in our highly regimented and structured societies.

Sensitivity among people with ADHD is fascinating, important, and markedly different from that seen in HSP or autism. I think of sensitivity within ADHD as having two parts. First, there is a deep curiosity about and sensitivity to new information and stimuli, an experience not too different from that of a bee driven to discover all available pollen. Second, there is the sensitivity that results from being ADHD, especially if it's been unknown, where people become sensitive to criticism and being judged. It's hard to do well at some times and then at other times feel like a total failure—for being late, missing an appointment, missing a deadline, getting dates or times confused, or other results of having a challenged prefrontal cortex and struggling executive functioning. There is also a sensitivity to ourselves—our own emotions, regulating those emotions, and not being so hard on ourselves. It's no surprise, then, that it is not uncommon for adults with ADHD to have meltdowns or "blowups"—like an adult tantrum. The sensory overload gets to be too much when someone is trying to hold together multiple threads of information or expectations in a neurotypical context.

I have struggled since I was a child with these meltdowns, but had no clue what they were. I would feel very sensitive to being judged or criticized and then find myself experiencing sudden bouts of anger or frustration. Much of the experience is a kind of accumulated trauma—after repeated attempts to do things the "right way" and not being able to finally "get" it, the person with ADHD eventually boils over with frustration. And other people are surprised or confused because everything that's happening is invisible and under the surface. It's a struggle that many people deal with for years before they discover what's really going on.

## Higher Education

"I probably lasted about three months into my postdoc before I couldn't focus anymore and it felt like I wasn't present throughout the week," my friend Stephanie tells me one day about her undiagnosed ADHD. "I would go home mentally and emotionally exhausted and spent most of my days off just spaced out. Then on the weekends I would have two days to myself, but I was just exhausted and things became even more difficult. Time was flying, and I was accomplishing absolutely nothing.

"In grad school there was structure and variation in the tasks—and it wasn't so much sitting at one table working on one computer doing the same thing over and over. I was dreading the end of school for that reason." Stephanie was anxious about her lack of self-direction and was so hyperfocused on her education that she never stopped to think of alternative directions for her life after graduation. "Why am I always so unproductive?" she asked her therapist one day. "Why am I always spacing out?" Her therapist suggested that perhaps she had ADHD, and she thought, "No way."

Stephanie was my classmate in graduate school at the Harvard School of Public Health before either of us learned about our neurological traits. We always got along well. She grew up in Oakland, California, the first daughter of Vietnamese refugees. She was an anxious child, always observing and noticing details that her peers and family members missed. She also came out as lesbian and generally describes herself as the black sheep of the family. What I notice while talking to her is the way she describes her feelings of sensory overwhelm. She had done well in school, and was now teaching at USC, but something wasn't right. It seemed like what the modern work structures and styles were requiring of her were not working for her body and mind.

Stephanie identifies as having ADHD and describes several sensory processing differences that dominate her daily life. "Jarring noise I find very upsetting," she tells me. "I go on information overload faster than sensory overload, which is why shopping malls are incredibly difficult for me." She finds choosing outfits exhausting, as well as shopping at grocery stores. "The shiny lights and the boxes and prices—I'll forget that I have a list and just go look at stuff and absorb information, like the patterns and colors and the way they're arranged. It's not upsetting—when I'm bombarded with sensory information, it's pleasurable and I enjoy it, but when I have to actually process all of it, I feel exhausted."

Stephanie also tells me she feels like she "dissociates" when she's alone in a room, as she'll forget about tasks that need to get done in the real world. "Living alone is difficult because I'll wake up in the morning and there's no human presence reminding me that the things in my head are not actually happening and I need to be here in this present physical world doing things. There are days when it will take me hours to get up and brush my teeth. I'll just be sitting in bed doing the laundry and running errands *in my head*, but not actually *doing* them."

Stephanie holds degrees from USC, Harvard, and Johns Hopkins. She is highly accomplished and driven, not someone who "failed to launch." And what she describes as withdrawing is actually her sensitivity taking over—an experience often reported by women with ADHD. Many women I interviewed with ADHD feel incredibly fragile and "overly sensitive" to the highly regimented structures of their lives and report a sense of shame at not being able to comply with expectations. They can do quantum physics, write PhD dissertations, or travel the world as stand-up comedians, but when it comes to what is expected as "the basic duties" of being a "functional adult," they are left feeling deeply incapable.

Maria Yagoda was a student at Yale when she was diagnosed with ADHD. A talented writer and journalist, she later "came out" about her diagnosis and became a champion for other women with ADHD. "I will definitely get sucked into something and have to devote all my time and energy into that," she told me in an interview once. She also describes sensations of sensory burnout, anxiety, and sometimes overall depression. She can handle multiple and rapidly changing deadlines, but keeping track of her keys and wallet day in and day out is much harder. "Sometimes on days that are the craziest—different news stories breaking, too many meetings, family drama—I'm able to focus more intensely than I could on a normal day. I feel like I kick into this special productivity gear."

## Wake-Up Call

A contributing factor to all of this dissonance is that girls and women are neglected to a degree in the scientific research on ADHD and other neurodivergences such that teachers, doctors, and others are not able to accurately recognize these behaviors or thought patterns as relevant to ADHD or Asperger's. Since the majority of what we know about such neurodivergences is based on research on males, it's likely that thousands of girls and women are walking around unnecessarily miserable because of lack of awareness and misdiagnosis.

At the same time, if such neurodivergence is simply part of the human experience, as the neurodiversity framework would tell us, what do we do with all these labels? We need to recognize them as useful starting places to help people articulate and sort out and define experiences; but stereotypes get in the way and are far from representing reality, especially for women, and they describe only a fraction of the neurodivergent population.

With new research and anecdotal evidence comes the recognition that divergent neurologies are very much alive in girls and women and that such individuals are essentially hiding in plain sight. For example, a young girl or woman with ADHD is much more likely to have the inattentive type of ADHD and be prone to daydreaming, but also hyperfocusing—that is, have the ability to zero in very intently on one thing at a time, to the point of mastery. With her ability to focus on her books, assignments, and tests, she will often excel in school, so parents, teachers, or doctors would never consider ADHD. But once that young woman enters college or comes up against a significant transition in her life whereby routine structures are taken away—and she needs to depend on her executive functioning to navigate the logistical details of daily life that she never had to think about before—her experience changes dramatically.

Likewise with Asperger's and the autism spectrum: research and numerous interviews show that because of the way girls and women are socialized, they often mirror and mimic the behaviors of other females around them and learn how to "be" and interact with others. Internally, however, they put an extraordinary amount of energy into how to act—what to say, when to say what, how to move their bodies in a particular social situation. At some point in their lives, often when adult responsibilities become too much, the amount of energy required to continue pretending, or "passing," simply becomes too much. Depression, burnout, fatigue, and anxiety frequently surface. Women are often unaware that such pretending and mimicking are even happening. But as they need more mental energy in more and more areas of life, they begin to notice that it's harder and harder to keep up in former social situations or relationships. It becomes clear that there has been a long-standing disconnect between their *actual* inclinations and the ways in which they have acted out of social obligation.

Isabel was drawn to The Neurodiversity Project gatherings and told me that she was popular and at the top of her class until she went to college, and everything changed. A gifted child, she grew up in an open-minded family of artists and outside-the-box thinkers and never felt different or disabled. But in college she found herself overwhelmed with the fast pace and competitive nature of her peers and ultimately moved back home to finish college elsewhere.

Fast-forward to when she later married and had children and found she could barely function. She became sick and overwhelmed. Then, her son was diagnosed with Asperger's and ADHD, and she finally recognized her own neurodivergences. Masking had taken a huge toll on her emotional and physical life, but once the mask came off, her well-being improved dramatically.

"I think if I had read more about Asperger's up front, I would have been more accepting about the whole thing. It really does fit us to a T," Isabel tells me, referring to herself and her son. "One of the things that stumped me was that I'm not a computer nerd or a techie person or into coding or any of the stereotypical stuff that you hear about." Like many other women in her shoes, Isabel found solace in online communities—where neurodivergent women take to Twitter and blogs to share openly about their experiences. Indeed, such sharing has been a (literal) lifesaver for many. "Talking about this with other women and having that camaraderie has been really healing and invigorating, and I now regularly feel joy and awe. I have a sense of pride and honor."

Isabel has found a calling in creating art that reflects her reality. "It finally became clear to me that I was going to do art and that I was going to use that art to be a bridge between neurotypicals and neurodivergents. I think our natural role as those in the minority, at the fringe, is to be the mirror to society. We're big thinkers and creators, and we serve through our innovations and works of aesthetic beauty. That's what I feel my work is really about—maintaining that bridge and that sense of openness and understanding, which helps foster well-being." Isabel now holds art discussions in her home, loves photography, and is able to tend to her neurodivergent family's needs.

Denise is a medical student who studied English literature at Columbia University; she considers herself HSP, was diagnosed autistic in her late twenties, and her mother and sister are both ADHD. "My whole family is sensitive—in different ways," Denise tells me. "My sensitivity wasn't noticed because we all were [sensitive] in different ways. If I had been a boy, I definitely would have been diagnosed as something, probably Asperger's." Denise says she was a "toe-walker" as a little girl and was often in her own world, constantly reading and climbing trees. She says that in middle school her main friend banned her from taking books to school, which helped her to be more social. "I knew that I thought differently from other people. I've always spent a lot of time thinking about how other people think." Denise also found herself over the years in very long-term relationships, which she describes as similar to having a "seeing-eye human." Having a companion who wishes to go out into the world more often than she does encourages Denise to interact and socialize. "But if I do too much in a day, like go to a party, the next day the only thing I can do is sleep until that veil goes away."

The "veil" is how Denise names what protected her throughout childhood and early adulthood. Like Stephanie, Denise describes an experience of dissociation that dominated much of her early life, until one day it began to dissipate. "I started feeling more clear about things at one point after college. I was standing on a corner in Louisville, Kentucky, one day, and the fog lifted. 'This must be what other people feel on a regular basis,' I thought to myself. I realized that I'd lived partially dissociated for twenty-five years and then slowly started to come back into my body." Now Denise feels more empowered with choice about her sensitivities. On a practical level, as a medical student, she has had to ask for accommodations at school because she cannot spend hours on end in large lecture halls—something that wasn't an issue during her undergraduate years because the classes were much smaller.

Denise also hasn't suffered from some of the more debilitating aspects of going "undiagnosed"—she never had a crisis of depression or identity. Rather, she says, "My crisis point was the physical symptoms of joint pain, migraines, and brain fog. My ability to process was just done. There was a lot of confusion." The combination of new research about neurodivergent women, finding a like-minded community, and being in a mental health training program has helped Denise feel more clear in general. "I've spent the past two years in counseling—not because I'm depressed or anxious—but to understand things, to have choice," she says. "I think being an HSP is another way to look at the world. I was so relieved to finally have words for this and to know that this is a shared experience."

Denise has many gifts. Obviously "book smart," she has also homed in on how her unique internal wiring makes her an exceptional listener and health-care provider. "I experience somatic empathy a lot," she says. But she still struggles to explain to other people some of her experiences, given that scientific research is just beginning to take women like her into consideration. What are we to do when our very real experiences in the world—especially as women—have no corresponding representation in the scientific, medical, health, and neuroscience research? With brain scans of HSPs, the rise of research on mirror neurons, and prominent individuals like Sara Seager and Joel Salinas stepping forward—and, most important, with the long-awaited incorporation of female interest, inquiry, and representation in scientific research—we are at a point now where our collective experiences are bulldozing through a patriarchal, neurotypical agenda. Smart, talented, capable people are "coming out" and radically revising narratives about divergent identity.

## Chapter 4

# Sensory Processing "Disorder"

I n my interview with SPD writer and activist Rachel Schneider, she tells me that her SPD is simultaneously a difference to be celebrated and a disability to be managed. Like many other people with neurodivergences, she has good days and bad days. She's well aware of the gifts her SPD brings—she is sensitive, aware, and in tune with her surroundings—but she also feels the impact of not being able to work in particular settings, go certain places, or be in highly stimulating environments.

The SPD label mirrors the descriptions of the other neurodivergences discussed in this book, but its unique point is in its particular *physical manifestations. While Schneider shares the same emotional sensitivity, perceptive qualities, and proclivity for overstimulation that people with other neurodivergences experience, she and others with SPD also find themselves craving particular sensory input and strongly avoiding others. This is most often seen with fabrics, tastes, sounds, and smells. Schneider craves strong, tight hugs but steers clear of surprising loud noises. She will heartily engage with a close friend but will avoid too many encounters in one day and has a strict rule of working only from home, which she has had to negotiate up front in job interviews.*

*When I first spoke with Schneider, she was six months into parenthood with a new baby daughter. Confident and with years of SPD awareness behind her, she didn't seem too phased by the noise, routine change, and commotion of having a baby in the house and in her life. But that's because she had already begun to make accommodations for herself, and all her close friends and family were educated about her SPD. Like CC Hart, the massage therapist with synesthesia who avoids highways because she sometimes can't track where moving objects are in relation to her, Schneider has a similar experience of feeling overwhelmed in certain environments, and her husband and close friends can now pick up on her behaviors when she's starting to shut down or "lose her senses."*

*Medication for co-occurring anxiety has been helpful for Schneider, as has working with a psychotherapist and occupational therapist. Originally told by a doctor that she had "panic disorder," it wasn't until she was twenty-seven years old that her sensory symptoms were finally recognized as SPD. A psychologist suspected the diagnosis but referred her to an occupational therapist to confirm. Now Schneider says going to the gym is her "best medicine of all," and she feels much more connected to her body.*

## The SPD Experience

Although it is not yet included in the DSM, SPD has been taken up by psychologists, researchers, parents, occupational therapists, and numerous community advocates. The leading research, practice, and advocacy organization for SPD is the STAR Institute for Sensory Processing Disorder in Colorado, founded by Lucy Jane Miller, PhD. The institute designates the following key statements as indications of possible SPD:

- I am over-sensitive to environmental stimulation; I do not like being touched.
- I avoid visually stimulating environments and/or I am sensitive to sounds.
- I often feel lethargic and slow in starting my day.
- I often begin new tasks simultaneously and leave many of them uncompleted.
- I use an inappropriate amount of force when handling objects.
- I often bump into things or develop bruises that I cannot recall.
- I have difficulty learning new motor tasks or sequencing steps of a task.
- I need physical activities to help me maintain my focus throughout the day.
- I have difficulty staying focused at work and in meetings.
- I misinterpret questions and requests, requiring more clarification than usual.
- I have difficulty reading, especially aloud.
- My speech lacks fluency, I stumble over words.

- I must read material several times to absorb the content.
- I have trouble forming thoughts and ideas in oral presentations.
- I have trouble thinking up ideas for essays or written tasks at school.

Many women find themselves seeking answers for their heightened anxiety after they move in with a partner or have children and discover a significant difficulty with the amount of touch resulting from cohabiting. Other women have had challenges with balance, coordination, sensitivity to smells, resistance to sex not explained by other factors, and managing work tasks. Occupational therapists in particular have taken up the task to educate the public about basic senses such as proprioception (the sense of self and body position), interoception (the internal sense, including feeling warm or hungry), and the vestibular sense of balance and movement.

Much of the focus within the SPD community has been on getting SPD officially recognized in the DSM so that children can get accommodations at school and get support for occupational therapy through insurance and the medical system. This tension between demanding official recognition and access to support on one hand and the seemingly opposite desire of wanting to be respected for neurological difference on the other hand is common across neurodivergences and in the conversations that make up the neurodiversity field. It's important to note that both desires are equally valid and can coexist. Many people hold both understandings simultaneously within themselves. For instance, the statement "I must read material several times to absorb the content" can be understood as meriting treatment or accommodation but can also elicit pride in identity—such as someone who identifies as HSP needing repetition because she processes information very deeply.

The SPD label has come to dominate the practice of occupational therapy for children, but less so for adults, as it is often missed because of its overlap with autism and ADHD. Originally described as "sensory integration dysfunction" by A. Jean Ayres in the 1970s, SPD has since been taken in by the field of occupational therapy as an increased number of children present with sensory processing differences. I say "differences" because the descriptions of families who seek help for SPD make it clear that how people process sensory input varies widely. The common cluster of "symptoms" described within SPD primarily includes overreactivity or underresponsivity to stimuli such as touch or noise. In both cases, the polarized response can interfere with how a child or adult interacts with others.

Three subtypes of SPD are generally described: sensory modulation disorder (SMD), sensory discrimination disorder (SDD), and sensory-based motor disorder (SBMD). Within these subtypes exist degrees of sensory overresponsivity, auditory sensitivity, postural challenges, and other factors. This breakdown of subtypes was first put forth by Lucy Jane Miller and colleagues in 2007 as a way to solidify the parameters of a potential diagnostic category in the DSM. Some medical scientists have resisted including SPD as a "disorder," but many parents and practitioners want SPD to be properly acknowledged because of the effect it has on children, adults, and families.

This is similar to the situation with other types of neurodivergences. There is a sense that a particular neurodivergence does not make people inherently disabled, but they feel disabled because of the generally overstimulating environments of dominant neurotypical culture and settings. Autistic scholar and writer Nick Walker and others call this the social model of disability, as opposed to the medical model of disability. As we have already pondered, Who is disabled? Who is "disordered"? Who defines "normal"?

## Intimate Lives of Women with SPD

Lisa has challenges with sensory modulation, overresponsivity, and sensory craving. Everything is intense for her, and she gets overwhelmed quickly but simultaneously craves input, such as particular smells. She also has dyspraxia, which means she has trouble planning new movements with her body and sometimes bumps into things. She is creative, warm, and engaging, having worked as a school art teacher for many years. Now forty-eight years old, she has three children in their teens and twenties.

"When I become overstimulated, I have a meltdown—I start crying and can't stop," she tells me. "And this would happen several times per week." What follows surfaced in almost every interview I conducted about SPD: "I have trouble with touch. When someone touches me lightly, it physically hurts. My husband and I have had challenges with sex. That's why we went to couples therapy; we were trying to figure it out." It was actually Lisa's couples counselor who initially suggested SPD, but it would take time for things to become clear.

"I went to my primary doctor, and she didn't know about SPD or where to send me, but suggested seeing a psychiatrist." The psychiatrist, however, didn't know what SPD was. "I started crying in her office and ended up explaining it to her and left after ten minutes," Lisa says. "I didn't know where to go. She suggested looking at the big Chicago hospitals, but none had any info about SPD. Then I went online and the STAR Institute in Colorado popped up and was full of good information. I decided to go to Denver to see them."

This is a common story—women interfacing with doctors and therapists who have outdated notions of behavioral presentations that unknowingly mask hidden sensory challenges. With little education, training, or awareness of these issues among medical professionals and with stigma and stereotypes surrounding women and their "hysterical" experiences, women often have to be their own advocates and cycle through several doctors or therapists. "I've always had sensory processing disorder, and I never knew that and never knew what it was," Lisa says. "I was diagnosed as an adult and wasn't even aware that it was a thing. It was so spot-on. And it was really nice to know a name for what was going on."

Jen also found her way to STAR. "I had no idea I had SPD, but my son was struggling with his development, so he was in early-intervention occupational therapy from the age of eighteen months. I was filling out all these forms for my son, and I was like, 'Doesn't *everybody* feel like this?' And my husband was like, 'No.'" Formerly in sales and the construction industry, Jen was always at ease moving around multiple construction sites in a day but could never sit still at a desk. Now as a parent, she says, "Somebody's always touching me," which is challenging for her. "I had faced these same kinds of challenges my whole life. I had to have all my clothes extremely tight and had my sisters count and make sure all my buttons were buttoned up."

Jen often found herself saying things like "Everything's too loud" or "Honey, please don't touch me." "You mean not everyone feels dizzy when they walk into grocery stores?" she asks rhetorically. "I started getting occupational therapy and was very grateful. I do better just talking through my feelings with my therapist, recognizing and walking through what in my environment is causing me to feel a certain way. I don't need a ball pit or a swing, though I do love tight hugs and squeezes," Jen says, referring to the diverse ways in which occupational therapists work with clients, especially adults.

Both Lisa and Jen worked with the same occupational therapists at STAR. And as with Rachel Schneider, Jen found that physical exercise is a key component of her therapy. "If I don't exercise, I tell my husband, I'm going to the dark place." Exercise—in Jen's case weight lifting—helps prevent both anxious and depressive symptoms for Jen.

"My occupational therapist also puts me into a hammock—and that swinging motion gives me clarity," she says. "All of a sudden I'm able to realize certain things about my day or why I was feeling a certain way. But really, she's just helping me make sense of my sensory experience and helping me not feel bad on an emotional level. This is not only a sensory challenge, but emotional as well. I can manage the physical—I know I like clothes a certain way and I don't want to be touched. We do family air hugs."

## Women Engineering Change

Sarah Norris and Carrie Einck are occupational therapists at the STAR Institute, and they helped expand the adult and adolescent practice area at STAR, working with Lisa and Jen and others with SPD. They grew the program in 2016, which now includes a broad spectrum of work to help others become empowered and independent in daily functioning. Sensory work is just one of the specialties of occupational therapists; STAR, in particular, also does research, education, and treatment.

"A lot of information and treatments are out there for kids, but what's available for adults is very limited," Norris tells me. She and Einck have been doing extensive literature reviews, teaching and training other practitioners, and developing a program for adults and adolescents who are sensitive to crowds, bright lights, unexpected touch, and uncomfortable clothing, and who feel anxious. "It's not always about trauma in childhood," Norris says. "A lot of the women who reach out to us in adulthood have already sought out a lot of other services by the time they get here and have received many diagnoses—typically depression, anxiety, and bipolar. And none of those diagnoses ever seemed to really fit, and the treatments weren't working. They would fill out the checklists, and it would just not feel like a fit and didn't make sense. And then once they got their hands on sensory-based information or research, they have this moment of 'This is me! This makes sense!'"

It's not that other contributors or diagnoses can't coexist, adds Norris, but the sensory information provides a new frame that makes the picture more complete. "When I reflect on the whole life span, from infants with SPD to adults with SPD, a critical time is often in middle childhood and early teenage years—that's when the mental health layers begin to compound sensory challenges." Norris points out that many gifted and sensitive children develop severe anxiety, for example. And as they move through their teenage years into adulthood without proper support, they face multiple mental health diagnoses.

A typical client of Einck's is a woman who's newly married or a new mother and is noticing new challenges related to sharing space with other people—tolerating new sounds and experiencing an increase in touch and other sensory stimuli. Sometimes women realize their own sensory processing challenges after their children are diagnosed with one. Because of that, Norris says, "many of the parents are able to articulate in detail what their experiences are, and it had helped Carrie and me reflect on our work with children because now we have someone who can tell us what it feels like for themselves." Einck and Norris dream of a day when SPD and occupational therapy are more widely understood by the public. "A lot of people don't even know what occupational therapists do," Einck says.

## The Occupational Therapy Experience

Einck explains that each person receives individualized treatment through various experiences—it's not like popping a pill. A sensory integration clinic typically has a gym and an assessment area. Bright blue and red gymnastics mats are spread across the floor, swings dangle from the ceiling, and smaller tools are tucked in drawers and cupboards for finer exercises. An occupational therapist will begin with checklist forms to get an initial sense of a person's strengths and challenges; sensitivities to smell, taste, and sound are all assessed. After the assessment is feedback—offering the individual a way to make sense of what was observed and felt while trying the swing or balls, for example. Is there a vestibular sensitivity (that is, one associated with movement through space)? Is the person prone to becoming car sick? Then comes the goal-setting portion to help a person integrate and improve upon very practical, tangible parts of her day, from sexual intimacy and touch to managing fatigue or making grocery shopping tolerable. Another big area is managing sound sensitivity, especially when kids and babies are loud or crying in the house.

Einck interchangeably mentions function, joy, and calm; these are the primary goals of occupational therapy in her eyes, especially for work with adult women. Many women can't identify what calm feels like or what high arousal feels like, she says, but using particular fabrics and stretchy material or experimenting with heights and trampolines will all help give a sense of a woman's ideal level of stimulation. Feeling an over-all sense of "regulation" is what many sensitive neurodivergent women are after, as anxiety is so common.

Currently there's not a lot of uniformity in how adults with SPD are treated, so I believe this is actually an exciting time to seek out care. Women with SPD have the opportunity to ask for what they need and want, letting the occupational therapist respond directly to their needs and requests in the moment. Do you need more touch and pressure, or less? Do you find that certain colors are more soothing than others? It's almost like having someone respond in real time to your triggers and soothers.

Lisa, for example, who is Einck's client, was introduced to the Integrated Listening Systems (iLs), an intervention partly based on the research of Stephen Porges. Involving a headphone set with therapeutic music, the iLs and Porges's research have been lauded for their practical applications in soothing the vagal nerve (a nerve implicated in both post-traumatic stress disorder and SPD). She also has a ball, trampoline, weighted blanket, and brush for her skin, all introduced by STAR. By using such tools, Lisa gives her body the input it needs to feel soothed and regulated, whether she's jumping, bouncing, cozying up to certain fabrics, or feeling pressure applied to her skin. This is a primary function of an occupational therapist—to simply guide and observe as a client interacts with such materials and devices and to help point out when the client appears soothed or hesitant, or startled or excited, or uncomfortable. Many of us may unknowingly get triggered by everyday textures or physical requirements; the occupational therapist helps point out both the triggers and the soothers.

The iLs has had the biggest impact for Lisa, who says the sounds relax and calm her, but help her feel alert at the same time. She tells me she hears, thinks, sees, feels, and comprehends better. Her attention and executive function are much more online as a result. "Now it's very rare that I have a meltdown," she says.

When I ask Einck about emotional overwhelm and meltdowns, she explains that emotions are always overlayed with senses. Even neurologically, there is a "dual coding," such that anytime a person has a sensory experience, an emotional experience happens at the same time. "It's imperative to address emotions *and* sensations, because they're both happening and they're both real," Einck says. Lisa underscores this point to me. "The crying is a neurological reaction," she says of her previous meltdowns. "So it isn't caused by emotions; it's caused by the brain being overstimulated. It's the weirdest thing because in my mind I know I'm fine, but I can't control the crying. In my mind I tell myself I'm okay, but my brain has had so much input that I can't take it anymore."

Norris tells me about a client who was complaining of feeling overwhelmed and tired all the time. She spent a week at STAR doing an assessment and creating a "sensory lifestyle" based on her needs throughout the day, starting with waking up in the morning and easing into the day. Bouncing on a trampoline, brushing her skin, and chewing on a chew toy all became part of her routine. After six months she was feeling like a better teacher, partner, and mother, and she was finally feeling fulfilled by her daily activities.

The eventual goal is for psychologists, social workers, primary physicians, and others to understand the sensory lens and integrate sensory care into their practices and make referrals to occupational therapists when appropriate. But more awareness, education, and outreach are needed first. "The combination approach is so much more effective than doing one at a time or doing either/or," Norris says. Therefore, Einck always likes to make sure that a psychotherapist is being consulted simultaneously so that her clients get the best care and support, and ultimately the best plan is determined.

"I think our two professions working together is the future of adult SPD treatment," says Einck of occupational therapy and psychology. "You can't piece apart having a tactile experience from an emotion. The two areas of the brain are constantly interacting and constantly talking to each other." What often happens with the women she sees is that they immediately direct their attention to the emotions they're experiencing rather than the sensory experience. "But what is the sensation?" Einck asks them. After swinging on a swing, a client may say she feels scared, but taking a moment to observe accompanying dizziness will reveal a lot about the woman's underlying sensitivity—in this case a vestibular sensitivity.

Occupational therapists like Einck and Norris help people understand their bodies and their reactions to stimuli. Pain, pleasure, and neutral sensory experiences are dual coded with memories. Grandparents, smells, colors, difficult times, lakes, valleys . . . sensory cues hold clues to our past and our memories. Many women latch onto language from popular psychology, such as "panic attack," when often they are instead experiencing sensory overwhelm. The treatment for a panic attack may not work for the adult woman who in fact has sensory processing differences. Frames of reference matter, and consequences of a misdiagnosis or misattuned therapy can be very serious. A woman may sit in a therapist's office for years looking for answers in her childhood; but again, Norris reminds that there doesn't always have to be severe childhood trauma, as Freudian psychology would tell us. Sometimes there is, but often there isn't. Sensory overload can look like anxiety, but once women have more information, they can better share and explore the sensory angle with a psychologist or other therapist.

## Occupational Therapist as Rebel Leader

"Historically, for a long time there was a belief that the problems that these individuals were showing in areas of sensory and motor challenges were not things that carried over into adulthood," the executive director of the Spiral Foundation (Sensory Processing Institute for Research and Learning), Teresa May-Benson, tells me. "Up until the 1990s, people believed that kids grew out of these issues. There wasn't a belief or understanding that adults could even have these sensory or motor issues. In recent years research has been looking at this, and we know that adults do demonstrate sensory processing problems."

May-Benson impresses me from the minute we start chatting; she articulates thoughts I've had as a journalist, but because she holds a doctorate in occupational therapy, she has hands-on experience with adult women, and case histories pour out of her during our conversation. "Part of what happens is that some issues in children that are sensory related—in adults they become psychological problems," May-Benson says. "And the adults end up being seen as a psychological issue. Here [at the Spiral Foundation] we see a lot of adults who've been down the road with the psychological community and been given a whole string of diagnoses from psychologists, and nothing has worked for them . . . and when they finally come here and we tell them they have sensory defensiveness or praxis problems and that's why they can't get organized, they feel immensely relieved and say, 'Oh my God, someone finally understands what I've been going through!' "

The invisibility comes with autonomy, May-Benson tells me, explaining that adults simply avoid situations that make them uncomfortable, whereas children can't. An adult may not know why she avoids particular situations, but she does it anyway. Children invariably wind up in situations that confront their sensory challenges—such as noisy playgrounds, an annoying toy pushed in their face, or loud siblings. For adults who avoid certain noises or locations, this may impact their lives in different ways—they go on fewer social outings or have a lower tolerance for touch in romantic settings. Then the focus becomes behavior, and they seek out a psychologist. This is what happened with Schneider, who later became an outspoken advocate for the SPD community. The signs were there during her childhood, but no therapist or doctor understood or correctly identified her sensory challenges. All they saw was anxiety. This is common, says May-Benson, and creates a difficult position for adults. You may hear someone say, "She's my clumsy friend" or "She's a picky eater." Or the most common, "She's just sensitive."

Many parents quickly refer their children for any sign of difference in eating behavior or fussiness with clothing, and so

SPD is more often recognized in children. "Parents are more likely to notice and want to address issues," May-Benson says, "so sensory integration intervention and assessment have been focused largely on children." Even for adults, occupational therapists are often trained to work only with elderly populations in nursing homes, not the general public with sensory challenges.

Since SPD is commonly seen in autistic individuals, May-Benson says that "there's been more awareness of these sensory issues, at least with autism," as children with autism are growing up. As a result of this growing adult autistic population and the sensory challenges that come along with autistic adulthood, researchers and practitioners have started to accept that adults more broadly may have sensory sensitivities and that they are not exclusive to autistic adults. "We've known this for years and years," May-Benson says, "because when parents come in, eighty percent of the time they say, 'Oh! Just like me!' So we've already seen this in parents. We know these sensitivities persist into adulthood."

When women come in to see May-Benson, her team relies on a suite of tools, including an adult/adolescent sensory history (AASH), psychometric research, and interview questions about lived experiences. "We tend to find more adult women seeking services than men"; the men who do end up walking through her doors often have previously tried to self-medicate sensory overwhelm with alcohol or drugs, she says. But many women come to her seeking support for being "sensory defensive" or because they have difficulty keeping a job given how disorganized they are. "We see a huge amount of anxiety. People who are very sensory defensive, super sensitive to sensory information, are being bombarded their whole lives with sensory input that is hard to manage—so essentially they are being exposed to ongoing trauma. *The disorder itself is actually just life—it's traumatizing them.* So those individuals are likely to have higher rates of anxiety because they've learned to live in a world where everything is making them anxious. Also related to that is depression, since anxiety and depression are interrelated."

## Sound Sensitivity and Misophonia

"The symptoms for me showed up when I was seven or eight years old, and I'm thirty-six now," actress Kathryn Renée Thomas tells me. "I have this memory of being suddenly very, very annoyed by the sounds of my parents chewing at the dinner table. Suddenly they had to deal with this nightmare of a child who was constantly getting angry at them." Thomas is the star of *Teachers*, a television show she helped create along with her comedy troupe from Chicago. She makes me laugh and I feel at home with her because her voice sounds almost exactly like that of my best friend from high school, who was also a comedic actress. "I didn't have symptoms before then, although I did have tactile sensitivity when I was preschool age. Getting me dressed was always a nightmare because I didn't like the way certain socks felt or the elastic on my pants. And my wristband had to sit on this exact spot," she says. Thomas is describing a child with SPD, though she didn't have a name for it then.

But what has dominated her adult life is a specific category of sound sensitivity called *misophonia*. While not officially part of the SPD diagnosis, many SPD researchers and advocates relate the two. Misophonia is sensitivity to sound, often severe, and the sensitivity to chewing sounds is common among misophonians.

"I've had it my whole life since then, and I definitely experience anxiety," Thomas says. "I get fidgety, have a racing heart, sweaty palms, and anger is for sure the way that it manifests itself. I get really, really angry. So whatever is the source of the noise, generally a person, I end up giving them a lot of dirty looks. We're shooting a television show right now, and in between takes the other day one of our camera operators was chewing gum, and he snapped it. It made me jump and turn around and want to throw daggers at him. I definitely get angry. I can't focus. It just throws my focus out of the room."

Gum tends to be her number one trigger, and sometimes she wears headphones because she can't focus otherwise. When she's in situations she can't avoid, like being in the writers room for her show, she tries to focus elsewhere and divert her attention— but that separates her from the conversation, and sometimes she feels behind or lost because she's been in her own world trying to escape the trigger.

Emotionally, Thomas says, she feels guilt. The women she started her comedy troupe with are like sisters to her, and she feels bad because sometimes when her anger comes out, it feels irrational. The same happens with her parents. "I don't want to give my loved ones evil looks, and I end up feeling really guilty."

"I didn't know it had a name until three years ago," Thomas tells me. "I spent more than twenty years thinking I was just kind of weird and it was a pet peeve, but clearly this was not a pet peeve." One day when she posted on Facebook a comment about her extreme annoyance after a meeting in the writers room, someone suggested she might be experiencing misophonia. She immediately looked it up and was blown away to discover that other people experienced it too. She then began opening up about it with her husband, parents, and coworkers. She didn't want to seem like she was searching for an excuse, but she wanted people to be aware.

"I also experience visual triggers," Thomas tells me, so even seeing someone's mouth move can trigger her. She has memorized people's chewing styles and names her friends who chew like bunnies or cows. I admit our conversation is hilarious—she is a comedy actress after all—but I know this is a serious neurological trait. "I'll make sure that the people sitting next to me block whoever annoys me." Her dad, who has a mustache, can even enrage Thomas when he twirls his mustache. "It's the repetitive movement that triggers me," she says. Her mom moving her fingers in a certain way on her cell phone also bothers her. As does fingernail clipping.

Thomas's "antennae" are always up, contributing to her success in her crowded profession of acting. What some may see as a disorder or weakness is the very thing differentiating her from so many others and driving much of her leadership as a writer and actor. She notices details, incorporates her awareness into her writing and theater, and produces comical work that can seem cathartic. "I tend to have a lot of anxiety in general, even outside of the misophonia. So I started doing yoga, which I really loved, *but there was always a person who breathed the loudest to show they're the best at yoga.* And that would always trigger me, so it just wasn't relaxing. That person did not need to breathe that loud!"

Thomas initially thought she was having what she called panic attacks; her anxiety has always been a standout feature for her. "I'm an open book," she tells me. "My mom is a therapist, and we were a family that was very open about mental health." She's had periods of depression, but she never clearly understood the extent of the anxiety. The medication Lexapro has been helpful for general anxiety, though she's not sure how much it has helped with misophonia.

Even Thomas's therapist had never heard of misophonia. "When I found out it had a name, I started looking up various doctors, but everything I was reading indicated how early it was in the research and discovery period. The majority of people were not having luck with seeing doctors and other types of providers." So she talks about it with people when necessary and she takes earplugs to places like movie theaters in order to not hear people chewing. "Kelly Ripa has misophonia, so it has to be real," she says, half-jokingly.

Thomas tells me about a newsletter called "Allergic to Sound" that has a large following and helps people to reframe what it means to have misophonia. She says she likes the analogy of an allergy because her misophonia reactions feel like an automatic response, just like an allergy. And on Facebook groups, she tells me, folks feel free to rant about "those damn loud apples" and other seemingly innocent offenders. Finding groups of like-minded people is encouraging and helps her see the humor in her experience. Countless other women have found such groups in person or online, and countless more sense they need such connection but don't yet know what they're seeking because sensitivities like misophonia are so underacknowledged in the media, let alone in the research.

## Coping with SPD

"Sweet kid! Sweet kid!" Spiral Foundation's Teresa May-Benson repeats to me over the phone. Then she tells me about a teenage boy who oscillated between getting caught for smoking pot and then breaking and entering. "The only time I feel totally calm is when I'm doing drugs," the boy told her one day. When May-Benson asked about the breaking and entering, he said, "Well, it's a thrill."

This kid had several sensory processing differences, and he partly dealt with them through substances and seeking thrills. May-Benson explains that girls and women, on the other hand, often may deal with sensory craving or feelings of overwhelm internally—and invisibly—by resorting to excessive worry; thoughts and emotions gradually become more challenging, but no one notices. Or they may read all day, turn to art, or in some cases say something to family members.

Gender socialization, gender norms, media stereotypes, and cultural attitudes have greatly affected the way that women and men seek out support for or deal with sensory experiences. In the case of women, many internal layers often build up in the form of anxiety and depression. In the case of men, however, coping mechanisms such as alcohol dependence are more common, May-Benson tells me. She continues: "This was a kid who was very tactile defensive, very sensory defensive, who felt he needed to do drugs—particularly pot—in order to calm down. But on the other hand he needed something that was challenging to his system and could get the adrenaline going. So the breaking and entering would give him a rush."

I know for myself in my twenties that I had a reliable habit of every few days needing to stir things up as a way to release pent-up energy. I would start to sense an anxiousness, a restlessness, a feeling of wanting to bust out. And invariably I would start arguing with my husband, as my default outlet has always been my voice. This conversation with May-Benson is revelatory.

Another of May-Benson's clients was a young man who was very underresponsive and had trouble with driving a car because he would lose his sense of spatial awareness; he wouldn't know where he was. And so he decided to ride a motorcycle instead because the vibrations, providing an intense sensory experience, helped him know where he was. "This is why so many young people seek things out like bungee jumping, because they can't get that kind of intense experience in everyday activities." People seek calm or seek intensity, she explains. There's a process by which all of us are seeking our sweet spot, our ideal point of regulation where we feel comfortable, in the zone, and in our flow. We take in basic sensory information from our environment; first, we assess its importance and relevance, and then we use it to regulate our arousal level so we can either habituate to inputs that aren't important or increase our attention to inputs that are threatening or relevant. If that process of modulation works a bit differently, then a person may have a high arousal level, remaining in a state known as fight, flight, or fright. That person's system is constantly being alerted to information that most people don't pay attention to.

Individuals who have challenges modulating information may constantly notice the buzzing of the refrigerator or the lights above, explains May-Benson, and therefore their arousal level is constantly being triggered. They might get irritable, be unhappy, or be unable to concentrate, so they end up having problems staying focused. Such individuals are antsy and on edge. May-Benson tells me about a woman who sometimes accidentally punched her boyfriend when he tapped her on the shoulder or tried to initiate a hug, leaving the woman filled with embarrassment. She often yelled as well—which I could relate to as I can sometimes yell when I feel startled. "The emotional reaction overrides the higher executive function in the moment," May-Benson says.

I tell her more about my personal challenges at home. Sometimes when my husband and I discuss logistics, I switch into battle mode. "We see women like yourself a lot," May-Benson says, "especially that overresponsivity and defensiveness. And in terms of the home life, that's often where the problems surface and cause the biggest issues."

The idea behind sensory integration work, May-Benson explains, is to help your nervous system adaptively respond to your environment. "The inputs"—referring to the toys, balls, swings, and other equipment typical of occupational therapy clinics—"set up your body for skills." This is soothing to my ears, and the gymnast in me is stoked. "We want you to be able to process this information better and more effectively, not just learn splinter skills," May-Benson says.

## A Learning Process

"All I can say is that as the days go on, I feel my sensory issues less and less," Rachel Schneider tells me one year after our initial conversation. I wanted to find out how she was doing. Her daughter, now a year and a half old, has earned the "toddler" label, which brings with it increased sensory demands on Schneider and her husband. It was clear that "mother" was becoming as much a part of her identity as advocate for the SPD community. "My daughter's needs come first, and they always have, and I think in some ways I'm exposing myself to things I normally wouldn't have exposed myself to," Schneider says. She's not trying to deny her SPD—just the opposite. After initial phases of diagnosis, acceptance, relief, and advocacy over many years, she has integrated all her learning and experience into her lifestyle and her sense of self. The experience or feeling of "other" has no place in her life now, and she has even taken her expertise into the workforce, becoming a diversity and inclusion expert for human resources professionals and recruiters.

"My husband and I took our daughter to music class a couple weeks ago, and I was worried, but I had the best time of my life. My daughter has shown me how much I can actually handle outside of the barriers that I've comfortably set up for myself," she explains. "Before I knew about SPD I spent much of my life being pushed over the edges of what I was comfortable doing. Nobody knew why I had a limit to where my comfort ended, but in adulthood I've found where my walls are and I know how far I can go. So having my daughter has given me space to push those farther than I thought possible in many ways." For example, Schneider tells me about a time when she and her sister took their babies to a kids' pool, and she had so much fun that she wasn't even thinking about her SPD or her sensitivities and usual limits. She's able to handle more than she expected, although, she says, "I'm also burned out more than I used to be, for sure."

The running and constant movement have been easier for her to tolerate but still provide a sensory challenge. Schneider says bath time is particularly hard, which surprises her. "There's a lot of visual and auditory hitting me at once"—the running water, the squeamish child, and the feeling of the water itself. "And of course there are times when I have a shutdown, where I have to go to the other room and lie down on the floor and take a deep breath."

I ask Schneider whether some of her surprising experience

with motherhood may stem from *sensory-seeking* inclinations. The music and pool time both strike me as forms of sensory stimulation that could be deeply pleasing and fulfilling, and many of us are unaware that we are actually craving these things. "I do find water peaceful," she says, "but the main thing is that this is for my daughter and she's happy."

Schneider also continues to pay attention to her needs. When her daughter was taking music classes, Schneider asked the class director about making a scheduling change that would allow her husband to be present more often, which helps Schneider feel calmer in such settings. The director praised her for sharing her story and asking for what she needed for her and her family. "I want my daughter to experience the world," Schneider says, "and I don't want my challenges to hold her back in figuring out who she is."

This is a theme of her life, says Schneider—encouraging people to do the things she feels she cannot do because she doesn't want to hold other people back. She knows that her sensitivities are often positive and unique, but she doesn't want to let the challenging parts, like meltdowns and shutdowns, affect what her daughter, friends, and family experience.

Regarding work, Schneider was just finishing a five-year work-from-home contract in copywriting and was in the process of interviewing in related fields such as branding and recruiting. She decided to disclose her SPD diagnosis to employers up front, because it's part of what she feels makes her unique. Through her work in human resources copywriting, she helps companies recruit individuals by helping those companies tell their stories— and this theme of personal difference and diversity runs through-out her own life and her work. So she wants to work with teams where she is welcome, and disclosure has been an important part of feeling happy and comfortable on a team. "Here's who I am and if this works for you, great; and if it doesn't, better to know now" is how she explains her approach to interviewing and trying to get hired for a new job.

"I'm very excited about differences," Schneider says. "I think they make for a unique tapestry, a richer culture." And she revels in the fact that so many people are starting to stand up and reveal their sensory differences. She encourages people to disclose, especially at work, as that makes for an inclusive workforce. "I get to write about a diverse and inclusive workforce, and I get to *be part of* a diverse and inclusive workforce, so it's neat to have both sides."

Regarding her life at home with her husband, Schneider says that he knows her sensory issues better than anyone— sometimes even better than she knows them herself. When she's overstretched, he sometimes gets the brunt of her exhaustion, but there's a deep understanding. She recalls a situation when she felt pushed to the brink and "barked" at her husband after seeing how calm he was in the next room. But, she says, they both knew it was because she hadn't had her quiet downtime or used her therapeutic brush (what's called a Wilbarger brush, which is usually run along the skin of the arms).

Schneider and I talk about how much "managing" we've had to do in our everyday lives—managing our boundaries, energy, sensory challenges, activity levels, relationships, work life, and more. For instance, if she's invited to go to the zoo, she immediately imagines the heat, sounds, animals, and crowds of people. But if she's invited to a swimming pool that she has visited several times before, that familiarity makes her feel more at ease. And for mothers, there's even more to manage. "I honestly believe that moms with sensory issues are superheroes. We are superheroes on a daily basis because of how we have to live our lives!"

Schneider was worried about being married and sharing a home with someone and then having a baby and dealing with sensory overload, but she's learned more about what she needs, makes sure she's around people who appreciate her, and finds she's able to be there for her family because she's so enthusiastic—and that motivates her. "I think the more we own ourselves and our differences, the more we become comfortable with ourselves and our differences."

# Part III

# SOMETHING NEW

Chapter 5

# Well-Being

**W**alking toward the front door of a local occupational therapy studio near my house, with my hoodie cozied over my head, I feel a bit in awe at the wonder of my path—from a research perspective and from a personal perspective. I had been a gymnast throughout childhood, and images of swinging around bars often dangled in my mind while I was on walks or long drives. Now here I am walking into familiar territory, with bright blue and red mats sprawled across a large warehouse-style room and swings made from shimmering fabrics hanging from the ceiling. A kind of "full circle" feeling washes over me.

Leigh, the owner of the studio, shakes my hand and shows me to a meeting room with child-size tables and chairs. (Occupational therapy offices focused on sensory integration rarely receive adult visitors.)

*Leigh explains the basics of occupational therapy to me before we hold a couple of sessions over the next few weeks. She teaches me about pushing against the wall for a deep pressure effect and puts me into a huge body sock so I can feel the weighted relief. I immediately go out and buy one. Notably, what I learn the most from my time with her is how well I respond to simply being given more information.*

*I found the same when I was first learning about somatic therapies—knowing the language of the nervous system helped me better imagine and comprehend the mechanisms by which my body and mind interact with my environment. It's as though the missing piece has been mere knowledge about my insides—my internal makeup—the life of my organs. I later learn in an interview with leading interoception researchers in the UK that one reason autistic women are more likely to be anxious is because of this very disconnect—not being able to detect one's own heartbeat, for example, and that merely being able to feel one's heartbeat after exercise can be a soothing and reassuring activity.*

*Ultimately, there is no one-size-fits-all approach to improving well-being, "wellness," or "therapies" for sensitive neurodivergent folks, but building a tapestry is important. In the remaining chapters we will take what we've learned in Parts I and II and apply insights to daily life and real-world situations in work, relationships, and wellness practices.*

## Altering Course

The question of where to go from here as regards to understanding and integrating neurodiversity thinking needs to begin with new insights into the nature of sensitivity and the impact of psychology on education, workplaces, family life, and more. It's crucial to understand where neurodiversity thinking fits within the larger context and evolution of psychology. We'll take a quick historical detour before meeting modern-day practitioners, researchers, and other pioneers and activists.

A primary resource to help shape our understanding of psychology's evolution is Duane and Sydney Schultz's *A History of Modern Psychology*, first published in 1969, which traces the origins of modern Western psychology. It designates 1879 as a particular turning point, when the German physician Wilhelm Wundt established the first laboratory dedicated to research in experimental psychology. Schultz and Schultz understand the history of psychological thought as movements that emerge within historical contexts, noting the importance of "contextual forces," such as intellectual, political, economic, and social factors that influence the currents of psychology. The authors also acknowledge the fluid nature of the field, noting that it is always changing and growing. A preoccupation with having a scientific basis for the field is apparent, as well as differentiating it from others, particularly the older discipline of philosophy. They note that understanding the human mind was once thought of as an intuitive exercise, but with the incorporation of methodical tools and inquiries into biological understandings of ourselves, that same approach could finally be transferred to the study of the human mind and human behavior. And so the history of psychology is largely the history of the evolution of such tools and methods.

## Mechanistic Thinking

The seventeenth century was the era of the machine. People began to think that if one could take apart a mechanism and understand its functioning, perhaps the same could be possible with the human brain. Philosophers of the time believed that "the harmony and order of the universe could now be explained in terms of the clock's regularity—which is built into the machine by the clockmaker just as the regularity of the universe was thought to be built into it by God." During this era, determinism and reductionism came to rule. *Determinism* is the idea that every consecutive act determines the next; *reductionism* is the idea and belief that an object (or person) of inquiry could be disassembled to ascertain exactly how it (or she) operated. Hence the scientific method was born, and the dominant "mechanistic" outlook within the sciences, including psychology, was that of humans as machines.

A host of influential thinkers—all white males—came to dominate psychological thinking, from René Descartes to John Locke, David Hume, John Stuart Mill, and others. One individual in particular, Johannes Müller, interjected a more physiological take on human behavior with his theory of the specific energies of nerves. This is in line with the medicalized notion of human behavior that we learned about in Chapter 1 in tracing the history of psychiatry. Schultz and Schultz write that Müller "proposed that the arousal or stimulation of a given nerve always gives rise to a characteristic sensation, because each sensory nerve has its own specific energy. This idea stimulated a great deal of research that sought to localize functions within the nervous system and to delimit sensory receptor mechanisms on the periphery of the organism."

Such a physiological take gave rise to a host of ideas, theories, and experimentations, including electrical stimulation of the brain and, later, lobotomies. The Berlin Physical Society was formed in the 1840s to explain to the world how mechanistic thinking could be applied to all sciences, including the mind.

Around this same time, insights about the nervous system, neurons, and brain function were being described in mechanistic terms, forming, along with empiricism, experimentation, and measurement, the backbone of modern science and psychology.

Around the turn of the twentieth century, Sigmund Freud developed his theories of the mind and psychoanalysis. His ideas were distinct and less reactionary compared to those of the earlier schools of psychological thought. He wasn't a clinical laboratory scientist or academic, and his focus on the unconscious made his ideas less prone to mechanistic thinking and less able to be studied using empirical inquiry. He favored inquiry into "catharsis," dreams, and sexual impulses. It's noteworthy that Freud regularly used cocaine and touted its "miracle" features until his middle age, when he was met with backlash and scorned for such assertions.

As psychiatry took shape as its own independent field within medicine, so too did psychology take shape as an independently respected field apart from what was earlier thought of as philosophical inquiry. In the late nineteenth century the first psychology journals appeared out of Europe and the United States. Laboratories started popping up, and the American Psychological Association (APA) was formed in 1892. The field became known as the "science of behavior" but remained low on funding rungs compared with other scientific disciplines.

With a population explosion at the turn of the twentieth century, psychologists turned their attention to universities and education, where funding was being directed. Psychology had been limited to clinical laboratory settings and was now growing outward into schools and the wider culture, so funding and prestige were finally coming its way. World Wars I and II would also invite more extensive involvement from psychologists.

Since Freud, a host of other thinkers and theories have made their mark—notably Carl Jung and Abraham Maslow, humanistic and cognitive psychology, and most recently what is referred to as "positive psychology." Spearheaded by Martin Seligman in the late 1990s, positive psychology aims for "optimal human flourishing" and focuses on enhancing people's strengths. It's worth noting that the same demographic that led psychology's history—white males—is also responsible for the growth of positive psychology, which, in combination with "mindfulness," has grown to dominate not only the field but also popular psychology books and blogs. It has been criticized, however, for resting on individualistic and capitalist notions of happiness and well-being and taking a uniquely Western perspective on personhood and notions of thriving.

A neurodiversity approach to psychology, however, does not seek to mute those aspects of ourselves deemed "less positive," unhelpful, or "less productive." Instead, as we have seen, the neurodiversity framework rests upon reframing the experience of humanity, particularly our notions of "disorder." Rather than trying to rid people of anxiety, neurodiversity proponents seek to uncover the source of the anxiety as having to do with how an experience of cognitive difference in our society creates that anxiety—and feelings of insecurity, alienation, loneliness, and depression.

## Sensitive Women

"Most people who come to me are highly sensitive and have been feeling misunderstood and not well heard by previous therapists," San Francisco Bay Area therapist Grace Malonai tells me. "I'm often people's eleventh or twelfth therapist that they've tried. They've been seeking support and help, so when they get to me they feel pretty raw as they are hurt and have been in need for a while." Malonai specializes in clients who experience high sensitivity across neurotypes, whether autistic, ADHD, HSP, or SPD.

"Things often come up at work or home" for her clients, Malonai says, "and they are not being well met or understood and they themselves have a hard time understanding why they are a certain way." They may wonder why no one else is so sensitive or sad at work. For instance, a colleague will say something to one of Malonai's clients, and the client will wonder what that person means and then begin to second-guess and third-guess and start to spin out into self-doubt. The client wonders and doubts because she doesn't see most people doing that sort of questioning. "But when clients come to understand themselves," Malonai says, "they begin to discern what information they are receiving is subtle or big as compared to what others are getting."

This last point is essential to how neurodivergent folks can be helped within a therapeutic neurodiversity-focused setting. Understanding and reframing difference becomes the key to shifting experience and relieving anxiety. Instead of your difference being what holds you back, its gifts can be uncovered. Some refer to this as unlocking a kind of "superpower," and discerning difference as power can be invigorating.

This begins with dismantling assumptions surrounding neurodivergent people. Stereotypes include the perception that such people are aloof or unaware of social cues. But in fact, as we have seen, what is actually happening is a hyperawareness of things that most people never notice. Malonai shares an example of a highly sensitive, gifted, and anxious woman who can become preoccupied if she sees her boss give a big sigh. The woman wonders about her boss, "Is she having a bad day? Does she have a stomach ache? Is she crying? Is she sad? Is she tired?" Then she wonders, "Did *I* do something?" Malonai explains, "There's a lot of doubt and a lot of insecurities."

But all that wondering can be a gift. I see this in my own life with my husband, who often calls me a "wounded healer," because I'll think I know how he's feeling when he says or does something, but I'll sometimes forget *to just ask him*. "So that way it's a gift, because you can use that information to connect," says Malonai. "On the other hand when you don't understand what's going on, you just think you did something wrong." She says that if we can begin to ask ourselves how to use this subtle information in a useful way, we may be able to avoid feeling overwhelmed. We don't need to react, but to notice.

I ask Malonai about "best practices" because so few practitioners focus on high sensitivity and neurodiversity. One that she practices is this: When people come to her after seeing a dozen other therapists, she reassures them that the subtleties they may observe from her in their sessions, such as a tired face, is simply that—she is tired. She doesn't want her clients to use up valuable energy, personal coping, worrying, and guessing trying to figure out what's happening with *her*. When she tells them this, "they are more at ease," she explains. It's a small gesture, but it's meaningful, and it's part of being a genuine and authentic human being. Normalizing their experience is important, because sensitive people are not getting feedback and validation within our culture. "I want them to know that it's okay to feel a lot. So many people have gotten the message that it's not okay to feel so much, emotionally or physically."

Giftedness is also part of our shared human neurodiversity,

Malonai says. The word *gifted* conjures up images of little Einsteins, but in fact the psychological study of "giftedness" has focused on a whole personality profile of gifted and talented individuals—and high sensitivity is often a part of it. There is also the term *asynchronous development,* which refers to exceptional growth and talent in some areas of life or study but delay in others. For some, this may be more along the lines of what's called "twice exceptional," where a person experiences both giftedness and autism, ADHD, or dyslexia, for example.

"I think part of giftedness is taking in a lot of information and then thinking about what to do with that information," Malonai says. "It's a form of asynchronous development," because the individuals are far ahead of their peers. "But they may be more developed in some areas and also less developed in some areas. And often less development has to do with social coping, and there can be overlap with high sensitivity, gifted, and autistic, so we want to help them understand their traits."

Regulating emotions is sometimes challenging, says Malonai. A person may say something negative, and the sensitive, gifted, or neurodivergent person will react defensively to the comment and may even scream or get into a rage. So then there's a behavior challenge, which makes a boss, partner, parent, or teacher respond with a negative consequence. "If you treat this behaviorally, you're not going to get to the root problem, which is *understanding,*" Malonai says. Intellectual explanations for why a misunderstanding has occurred or why feelings are hurt often work more successfully because there is a deep need and craving to understand the "why" of a situation.

On a practical level, I begin to wonder what it's like to be in Grace Malonai's office. Physical environments influence sensitive people in particular, and so I ask her what makes her working environment different or more comfortable for her clientele. Weighted blankets, fidget toys, soft materials, skin brushes, water beads, sand, pillows, electric candles, and other sensory accessories adorn her office. Also, the office has no overhead lights and strong-smelling foods are not allowed. Some of her adult clientele are visually sensitive, so she has set up her office carefully and intentionally. Solid doors keep noise out and quiet ventilation has been installed. Warm colors and clean interiors are important in creating a soothing atmosphere for her clients.

And what about the women who walk through her doors? Do they know they are autistic or HSP? Or are most on a journey of discovery? "I think when people come to me they realize they are different—some people know, some don't, some are misdiagnosed, some aren't. Some have been diagnosed as ADHD and autistic, when I might say they are actually highly sensitive and SPD, with accompanying anxiety. It could also be some amount of OCD. When we understand a diagnosis and we are able to treat the correct one, then the quality of life is increased."

The opposite is also true when women come to her thinking they are extremely sensitive but she sees a missing diagnosis of autism. "So we'll do an assessment and they will be bawling, saying, 'Oh my gosh, my whole life I was like this and I thought something was wrong with me. I didn't know.'" One woman in particular came to Malonai who struggled in college and had to stop going to school, but after getting the correct diagnosis of autism and requesting accommodations, she was able to finish her degree. Then there are other women who do fine in college, but they struggle when they get out into the work world.

Yet another group is serial job hoppers. With each career, they reach a level of success where they feel they've mastered what they want and then get bored and move on. But they think something is wrong with *them*. "What we've found is that they need to find a job or career that is not static," Malonai says, "or they need to have supplementary activities and hobbies. Or they have to be okay with the change and transition, but then there is the question of everyday life and paying bills." Many of these women are likely ADHD and gifted, often referred to as having "multipotentialities." Malonai points out that this kind of movement from job to job is more accepted in men, but she sees it all the time in women.

## Introversion and High Sensitivity

"I'm doing a lot of work with companies and schools and organizations these days, and I hear a lot of personal stories from people," Susan Cain tells me. "And some of the people who I talk to or work with are straight-up introverts, and then an awful lot would describe themselves as highly sensitive, so I'd say I work with both types." We are both parked in our cars during this conversation, and we giggle at the shared fact of loving to work from our cars because of the quiet and serene solitude they offer. Her debut book, *Quiet: The Power of Introverts in a World That Can't Stop Talking* (2012), was life-changing for me when I read it. I had also published my own initial "coming out" personal essay about neurodiversity on her website, Quiet Revolution, and now after months of emailing back and forth, we were finally catching up by phone. (Quiet Revolution has since evolved as a resource for educators, workplace leaders, and individuals to help them navigate introversion and leadership, and more. What Cain calls "introversion" she sees as a constellation of traits, often including high sensitivity and sometimes shyness.)

We all know by now that if workplaces, families, friends, bus drivers, rideshare service drivers, front desk people, and others knew about diverse temperaments, traits, and neurological makeups, then we might be far less anxious interacting with them and much better able to "perform." And I mean actually do stuff and get stuff done—without crashing, melting down, shutting down, or developing severe migraines. Most of the time we can't get every person in our ecosystem on board, and so it is partly up to us to figure out our own hacks, best practices, and other activities that help us navigate our days.

"It's helpful that people now have some self-awareness about these traits," Cain says. "We can't overestimate how important that is. A lot of people tell me this—that once people became aware of themselves as sensitive people or introverts or both, that gave them the permission to be themselves and carve out the space they needed to protect themselves. And it also gave them the confidence to use their gifts the way they were meant to be used." People are now emotionally giving themselves permission to *be*, she says. "And there's a weird paradox that happens—that the more people give themselves the permission, the more proficient they become in settings that you would think of as not suited to their best strengths." So, she says, the more people feel okay about being sensitive, the better they'll do in a stressful situation because they show up as their full self, instead of being full of self-doubt.

I'm nodding in agreement. "The keyword here is integration," I say. "Yes!" Cain responds, filling me with that good vibe that we are on the same wavelength. And then she explains how just being able to name something can be powerfully healing. "When I first started speaking to companies, they only wanted to hear about stuff that was relevant to people as workers or productivity or leadership. But now I've noticed they're really wanting to be helpful to their employees in a more whole-person type of way. Now they'll tell me, 'Spend time talking about raising introverted children.' I think there's more openness in general, especially in inclusion spaces."

Cain expresses delight at the way companies have started to open up about temperament being another form of diversity. And it's true—when such difference is respected and accepted, people feel more free to be themselves, and their best work comes out. Well-being and "wellness" are inextricably linked to work, productivity, and leadership. It's important that we understand ourselves and our own diverse bodies and temperaments so we can avoid outdated anxieties and thrive.

## Hearing My Body

Speaking of bodies, to give you some context, at the time of the following interview I had come to a personal realization that I crave a deeper bodily sense of who I am, where I am in space, and how I process people and the world around me—and how all of this affects my happiness, well-being, and relationships. I love heavy thick blankets, as many sensitive, neurodivergent people do. But more than that, spreading throughout my life— childbirth, friendship, love, grief, community—I had come to the realization that I crave processing events and experiences with my whole sense of body and self. Talking with a group of women isn't enough—I want to dance and drum alongside all the chatting. Exercise classes aren't enough—I want deep philosophical conversations as we are sweating it out in a gym. Being a journalist, I went straight to the experts and researchers who can explain some of this from a scientific standpoint.

Lisa Quadt is running a new clinical trial on improving heartbeat perception and measuring whether anxiety might be reduced through such perception. It's a simple yet powerful proposition. The trial is based on the research of her colleague Sarah Garfinkel at the University of Sussex.

People can measure their body signals in different ways, Quadt explains, such as sweating palms, feeling the heartbeat, or measuring the pulse. "We have found in autistic people, they tend to not be very accurate. But they think they are perceiving these signals all the time. Their perception gets overwhelmed with signals coming in all the time, so there's a mismatch." Quadt and her colleagues found out that the bigger the mismatch, the higher the anxiety—and the level of the mismatch predicts anxiety levels. "If we can match this—make the signals and perceptions more accurate, then maybe the anxiety will go down," she says. Again, the theme of anxiety is paramount.

So Quadt and Garfinkel are teaching study participants how to more accurately read heartbeat sensations—and monitor corresponding reductions in anxiety. This seems to echo the simple yet powerful realization I had been having around the importance of bodily awareness and how it relates to my feeling more centered in my daily life. It's as simple as counting heartbeats. Eventually Quadt and Garfinkel want to develop an app or have this kind of monitoring be a part of smart watches. She says certain types of mindfulness training could also adapt and incorporate interoception-focused therapy and heartbeat-counting.

They were still in the clinical trial phase when I spoke with Quadt, so she didn't yet know the final results, but in the meantime people can try timing themselves at home with and without a finger on their pulse and then compare the two results. This is easier to do if you do a couple of jumping jacks or other exercise just before you measure—if you stand up right now and try to merely sense your pulse, it will be tough. But if you stand up and do a few jumping jacks first and then try to sense your pulse, you may notice it's much easier.

When I ask Quadt about interoception in general and why it's perhaps more challenging for sensitive neurodivergent women, especially autistic women, she mentions sensory overload and high anxiety. Bus rides, for instance, present challenges with noise and dizziness. "The interoceptive signals that go from the body to the brain are kind of led in a different direction," she says. She adds that sometimes autistic and other neurodivergent individuals may notice their heartbeat and then overreact. Whereas a neurotypical person may simply notice it and then move on, a highly anxious autistic person may become alarmed and worried.

Cognitive behavioral therapy doesn't work for everyone, though I have found it somewhat helpful for reframing situations, behaviors, and interactions—and understanding my past. But after living in multiple cities, trying different talk therapists, and then trying more somatic ones—which did help with bodily awareness—it was clear that none of it would comprehensively address questions of autism, ADHD, sensory challenges, and more. That kind of improvement came from doing the research—reading studies, interviewing researchers, writing articles—and writing this book. As I have already said, *simply having information* can be life-changing—and one of the best sources of help, healing, and growth.

Something that has been working for me lately is imagining my amygdala—the part of my brain that helps process fear and anger—"cooling down," like shrinking or going from red-hot-fire red to a cool-calm-soothing blue. It really works! I always wanted to be one of those people who could find little hacks for myself, since so many mainstream approaches don't quite fit with me, and this has been one of my proudest breakthroughs. So when I start to get overstimulated or irritated, I imagine my brain, my brain stem, and the whole of my nervous system. I picture the entire system cooling down and the whole of my body filled with a velvety ocean blue. And I feel as though my body thanks me in the form of calming down. It's as though my internal body parts have been wanting my attention—*wanting to be acknowledged.* And once I acknowledge them, I can signal them, through vivid visualization and color, to calm down.

Not surprisingly, these intense bodily perceptions—of our own senses and those of others—are especially powerful in the lives of women. Quadt confirms much of what we have explored in this book already: women are usually diagnosed much later in life than men/boys, and the societal pressure to mask is much bigger for girls. Autism and other neurodivergent traits thus become expressed differently in men and women. And Quadt confirms what so many others have told me and I have related in this book: "For most women I see who got their diagnoses late, it was such a relief because finally they knew what was different about them and they could get information about it and meet other autistic women and see they're not so different after all and there's a community they can be in and just understand themselves better."

Quadt concludes our conversation by reiterating the heightened empathy autistic people have, as well as the need for society to make room for autistic and neurodivergent people. "I hope that the neurodiversity movement gets bigger and seen more," she says. "I don't see autism as a disorder at all. It's a difference in perception, emotion, cognition, and action, but it's just different, not lesser in any sense. I hope that the new research that shows high empathy, for example, will lead to societal change. We need to make space for everyone. And I hope research can help push that forward."

## The Other Side of Wellness: Medication

Medication is of course another route that many people consider when looking for ways to deal with heightened sensitivity and the impact it can have on their lives. I took Prozac (fluoxetine) for a year, and it seemed to dampen my sensitivities a bit, particularly the rageful feelings I had during logistical conversations with my husband. That was the only feature I really wanted some support with, since I love how my sensitivities inform the rest of my life, even if challenging at times. I tapered off the Prozac just as my career was picking up steam. (As a side note, a key personal observation I have been able to make is how central my professional life is to my healing. Knowing how I like to hyperfocus and become absorbed in meaningful material has allowed me to embrace that even more, and the energy of it has kept me in a steady place.)

Psychiatrist Lawrence Choy's story parallels mine in some ways. A UC Berkeley graduate who then went to Stanford Medical School, he didn't find out that he had ADHD until after he'd finished his psychiatry residency. He started digging further into new neuroscience studies and piecing things together, all the while getting treated for the ADHD. This new information drastically altered his perspective on psychiatry, and the DSM, and he launched an entire clinic in Silicon Valley devoted to entrepreneurs with ADHD.

"My approach is understanding symptoms and behaviors based on how our brain works, as opposed to what the DSM says," Choy says. "For example, with ADHD, if you meet a certain number of symptoms and depending on your age and functioning, you get the diagnosis. It's not based on neuroscience. But by looking at how the brain works and how the prefrontal cortex gets activated, we use medication to hack the functions of the brain."

He feels his approach is destigmatizing because he's not looking at ADHD as a disorder. But I noticed that the website for his clinic uses the term "disorder," so I asked him further about this. "Right now insurance compensates through the language of disorder. I wanted to make a radical shift, but I also have a business partner who reined me in and serves as a check and balance. We're operating between two models."

This is a challenge for many practicing physicians, therapists, and psychiatrists who are also neurodiversity advocates. There is a push and pull between diagnostic requirements and what the in-person experience is between patient and doctor. Straddling progressive approaches with current medical system requirements, Choy has learned to ride the waves. "I have ADHD myself, and when I went through this discovery, I was basically in this born-again mode where I was telling everyone about this. I realized I needed to start telling people about this new approach—over time, my intensity diminished and I learned that change takes time."

The current system and model treat symptoms and "disorders" with medications, Choy says, but they aren't really helping people get to the "next level" in their lives with meaningful goals and accomplishments. Medication can help a person focus, and once a patient is able to focus better, Choy looks at what the patient wants and helps that person "optimize" and move toward those goals and accomplishments. "Stimulants are still the mainstay of treatment [for ADHD]," Choy says, "but most clinicians use stimulants based on symptom improvement like being attentive and less forgetful. But on a macro scale it's about using these medications to develop the prefrontal cortex, which is where executive functions are"—such as attention regulation, impulse control, reality testing, and judgment. "If that part of the brain is more online, people can start seeing more things about themselves that they didn't see before." He says the more you use the brain, the stronger it becomes, "and then a transformation is made."

It's refreshing to chat with Choy. Language and vocabulary choices still echo those of pathologization, but he and others are trying to move beyond that paradigm.

## Expanding Definitions of Mental Health Care

Chris Cole is a therapist for people in addiction recovery who specializes in the intersection of bipolar disorder and an alternative framing of mental health challenges called "spiritual emergence." This views "mental illness" as a process whereby deeper artistic and intellectual gifts begin to stir and ultimately manifest in innovative contributions to society. Cole hosts a popular podcast called *Waking Up Bipolar*, and we spoke one morning after months of emailing back and forth.

"I got into working in mental health via wilderness therapy, coaching, and instructing in a variety of mental health settings," he tells me. A self-identified HSP, he has helped spur awareness and action in his local professional community around embracing the neurodiversity framework within therapy practices and educational settings. In particular, he says, "I appreciate the intersection of queerness and neurodiversity," as he identifies as gender queer.

"I got really curious about having a different emotional experience and the positivity of sensitivity." He tells me how he started to understand more about the HSP label through research articles and found it a positive reframing of sensitivity. "The positive side is that I'm very in touch with what it's like to be emotionally attuned to people," he says. "My barometer for the insight into another person's nervous system is a little more keyed up, which serves me in a therapy setting as someone who is trying to help people articulate an experience or emotion. And as a parent in a family setting, I feel much more able to respond because of that sensitivity. I can give my two boys the gift of reading their emotional landscape and the micro expressions of how they're experiencing the world."

Cole says there is a harder part, though, which is that the level of self-care and boundaries he needs to have in place are much greater. On a practical level, he has to carefully manage his schedule, sleep, and food, and he practices mindfulness meditation. The "negative" part, he says, is having to accommodate the fast-paced nature of our culture, which burns him out quickly. "My sensitivity affords me a great passion for helping people and creating transformation in the world, but that has to be balanced with the unplugging and settling back into what feels more natural for my nervous system."

I ask Cole how he works with highly sensitive clients, and he tells me about both the practical and the philosophical aspects. "Being able to really validate the person for their sensitivity is important. There's such a tendency to be pathologized for being sensitive, and this pathologization—because we are social creatures—makes it vital to have another person reflecting back the positivity and wisdom of the sensitivity. I'm not very prescriptive, and I'm really asking open-ended, curious, motivational questions so that the therapeutic experience is personalized to them." He also helps people navigate seeking care and services, such as psychiatric medication. He wants to educate and empower people in the medical system, since so many patients quickly become pathologized.

"For example, if someone is diagnosed bipolar and they see a doctor for help, many people don't have enough information about what [the diagnosis] really means. A diagnosis is just a list of symptoms, and if you meet those symptoms for a certain duration of time, then you qualify for the diagnosis. So that's the first thing people need to understand: it's an appraisal of symptoms." He then discusses the utility of a diagnosis and points out that it's only as effective as the treatment it helps provide. One of the reasons to have a diagnosis is to know what kind of treatment to give, he says. This information helps people orient to what is actually happening. "My job as an advocate for helping people to think about this through a neurodiversity lens is to ask the person to consider which symptoms are a problem. That restores agency. Symptoms like agitation or irritability are often named.

And then figure out steps toward basic harm reduction and improving quality of life."

Cole details the complex position he finds himself in at times when he is operating in traditional neurotypical psychology settings where pathologization is rampant. "I have compassion for folks who have power—parents and doctors, among others," he says. "I know many psychiatrists who are really big-hearted, well-meaning people." But he acknowledges the painful psychological symptoms that often arise for sensitive neurodivergent people in what he calls a "society that is so disconnected from its nature." Such people are expected to live their lives out in "such a callous world." He says it takes extra effort to operate in pathology-dominant settings. "It grates on me," he says. As a sensitive person, "I feel like I'm doing a lot more work than others in the room, because I'm holding the pain of the person who has this diagnosis and it's being discussed in a way that doesn't take into account the greater context of humanity. It's a kind of microaggression or subversive quality that takes a lot of effort to stay present with."

So then the question arises of how therapists and doctors at the cusp of neurodiversity-infused clinical care are to go about their work. As Joel Salinas said, the idea is to treat distress, not difference. And with any new approach or movement, there is a period of transition and the merging of new and old ideas. Practically speaking, says Cole, "In order to work with Medicaid, there has to be a medical jargon and treatment orientation in order to serve the person who needs services. So there's a challenge— I want people to have greater access to good psychotherapy, and at the same time in order to do that they have to find psychotherapy within a medically dominant paradigm. So it's a catch-22, because in order to abandon the medical system, the financial burden falls on the consumer. To pay out of pocket for therapy that is anti-oppression, that is hard. I want to serve, but I am also forced to diagnose."

So Cole says he has been taking charge of integrating neurodiversity with mental health approaches in counseling and is hopeful in his "little community" in Colorado, as he calls it. "Our community counseling centers are talking with me about my work around this. I have seen how counselors, therapists, social workers, and psychiatrists are very excited about the way that the neurodiversity paradigm could be implemented into mental health in a way that still allows people to get the services that they need. I think that's the growing edge."

The "edge" that is hard for clinicians to hold is how to still talk about the problematic or painful symptoms of such experience—and Cole says there has to be a new sophistication in how to hold the complexity of both the liberation of alleviating pain and suffering and the liberation that comes through the "anti-oppression of confronting the cultural, medical, and societal assumptions of these symptoms."

Since "trauma-informed care" has become a recent buzz-word, I also bring up the topic of trauma. The awareness of trauma and how it affects people's lives has been fundamental and important, but problems have sprung up—mainly a kind of mind-set that wants to attribute *everything* to trauma, as though some kind of "normal" exists that everyone would return to if they just resolved all their trauma or didn't experience any childhood hardship to begin with. This perspective is in danger of replicating the simplicity of past theoretical frameworks. There can be natural variation in the human species—as the neurodiversity framework suggests—and on top of that, some people may experience trauma. The integration of both viewpoints and frameworks is what is important and necessary.

## The Sound of Sensitivity

During the course of my interviews, especially with occupational therapists and those working in the trauma field, the topic of sound came up again and again. It may not be the first thing we think of when we think about sensitivity or trauma, but neurodivergent folks are affected more intensely by sound than neurotypical individuals. The role that hearing plays in our experience of the world cannot be understated.

At the heart of current thought and concern around sound is the effect that noise has on our nervous systems. Recent research into sound and well-being suggests that extreme low frequency sounds, like the hum of a refrigerator, and extreme high frequency sounds, like a car horn, trigger our nervous systems into fight-or-flight states. But what researchers and practitioners do with this information differs greatly.

Take Stephen Porges, a professor of psychiatry at the University of North Carolina, who developed a music-based intervention for people with auditory sensitivity, including those with SPD and ADHD, and those on the autism spectrum. The Safe and Sound Protocol (SSP) is an approach that helps steer people away from triggering sounds in order to rewire the nervous system and how it interacts with the middle ear structure. The SSP is based on research showing that establishing a sense of safety in the nervous system will result in calmer, more regulated behavior. The intervention involves a gradated music system heard through headphones that progressively sensitizes an individual to sounds—removing triggering sounds—and gets the person back to a state of regulation. Many adults and children have expressed intense satisfaction and overall reduced anxiety with the intervention.

Bill Davies, a professor of acoustics in the UK who is also autistic, would have us focus much more on the role of the environment, policy, and societal structures as opposed to individual interventions. He notes the role of noise pollution and new regulations emerging from governing bodies like the World Health Organization, for example. As an autistic person, he is much more concerned with inequality in sound design and thinking about how soundscapes affect people in unequal ways. He says cars, for example, are the biggest contributor to noise pollution. So his is a macro viewpoint on the policy of noise and how sound design can more intentionally help foster well-being.

Then there is Lindy Joffe, an occupational therapist who raises particular concerns about how schools are designed and built. For instance, they often rely on linoleum floors, exposing preschoolers to noise bombardment because of how sound bounces off such floors. She tells me about the importance of getting kids out into nature to be exposed to the sounds humans were designed to hear. Anxiety has become one of the biggest issues that people come to see her for—a newer development, she says. As for adults, much like the other occupational therapists I interviewed for this book, Joffe says that most who come to her for depression or anxiety have never had their underlying sensory challenges addressed. And so simple interventions in nature, such as brief camping trips, can be just as helpful for adults as children because of how natural sounds help regulate the nervous system.

So hearing, sound, and the middle ear structures are all points at which neurodiversity, occupational therapy, and the trauma field meet. Some may view auditory sensitivity as the result of trauma—which for certain cases it could be—but for many of us, the trauma comes from a world that is simply too loud. There are different kinds of trauma—the very nature of being human is to be subjected to life experiences that shape us, in both challenging and uplifting ways. Everyday life etches into our being, informing our personality and characteristics. As Teresa May-Benson said, everyday life is sometimes a trauma for folks with SPD, for example. So sound is important to think about in our daily lives and as we take steps toward improved well-being. Regardless of how it may come about, many of us just want a society and daily routine that feel more compatible with who we are and who we want to grow into being.

## Divergent Wellness: Tips for Taking Care of Yourself

You can take steps to begin making this shift into a sensory-compatible lifestyle. The following suggestions are taken directly from my own experience. Because I'm a journalist and not a therapist, I can't prescribe to you what you should do, but I can help synthesize what I've seen, observed, researched, and experienced.

- Take your time. This book may be your first foray into neurodiversity, or you may be a seasoned expert. You may feel ripe for change, or you may feel like you're just beginning to dip your toe into new waters. It's all okay. Let the ideas, insights, and realizations simmer.
- Try different approaches, and take what works from each. Don't feel pressure to stick with that one mindfulness meditation or that one group circle or other intervention or self-care tactic. Once you've soaked up

what you need to get from a particular approach, move on to the next.

- BUT, don't stop when something is working! If it's working, keep going. Feel free to move on when you've gotten all it can give (or when it's getting way too expensive).

- Educate yourself—in particular, fill the gaps in your visual knowledge and look up images, diagrams, graphs, illustrations, and more about the body, human anatomy, the nervous system, and the brain.

- Write down what works for you. If writing is cathartic for you or you have aspirations to write a book or publish an article, jotting down your notes can be helpful later in creating a narrative arc.

- Inform your family, friends, and in some cases your social media circles about your neurodivergence if it feels safe to do so. It can be incredibly healing to open up and share your truth. This is also helpful from a practical point of view, because your partner may want to know what you're up to at all those appointments (again, especially if they're getting expensive). It's also less lonely and alienating when your friends have some sense of what you are exploring so you can have conversations or at least check in about the process.

- Don't feel guilty when you start feeling better. Surprisingly, this is a hard one. It's so empowering when you finally land on what's been going on for so many years. You feel emboldened and refreshed (and of course sometimes angry, confused, or anxious). Then after a while, once the new information becomes integrated into your life and identity, it all becomes normal. And you feel fine. And maybe like fighting less. It's all okay. Integration is the point. Don't feel like you have to match a media stereotype of what a neurodivergent person should look like, which is more often than not miserable and uncomfortable. We are here to change that narrative, which means you need to boldly embody your neurodivergence—and it's tremendously helpful for the world to see you in your full joy and happiness.

## Chapter 6

# Home

Home, for me, is about environment and relationships. I have always felt deeply affected by my surroundings—feelings of sadness as a child when I was in sterile suburbs or office buildings, or feelings of joyful awe when I was by the ocean. And I think for sensitive neurodivergent folks, not only are we particularly affected by our environments, but we experience our environments differently to begin with because of our sensory sensitivities.

The same is true of our relationships—we experience the human dynamic differently because of our sensitivities. It is because of this that I would argue that our umvelt—a term from the German word Umwelt meaning "environment" and used to describe our sensory world—can be assumed to be experienced differently from neurotypical people as well. I think it's time that we take seriously the

*differing umvelts of autistic, SPD, HSP, synesthetic, and ADHD*
*folks as well—and create a world that we can all feel at home in.*

## Reframing Design for Sensitivity

A former designer at the design firm IDEO, author Ingrid Fetell Lee spent a decade researching what she calls the "aesthetics of joy." Through research with psychologists, designers, historians, and others, she came to see a clear link between one's environment and mental health and well-being. In her 2018 book *Joyful: The Surprising Power of Ordinary Things to Create Extraordinary Happiness*, she flips the script on what we think of as "overstimulating" versus "understimulating."

"If you think about nature as the baseline for what our senses are good at processing," she tells me, "nature isn't silent, quiet, or still—nature is always moving—and yet it's the most calming setting we have access to." So although we might discount loud noises or lots of movement as sources of calm, Lee questions such assumptions. "I think we mistake something that's calming for something that's less stimulating, when in fact I think a lot of our environments are understimulating." She points to a concrete dorm room or apartment, for instance, as being "a problem of understimulation."

Lee tells me about Snoezelen "multisensory environments," developed in the Netherlands, where autistic kids are exposed to cozy, soothing, darkened rooms with purple and blue patterned lighting, almost like a psychedelic room from the 1960s or the set of an Austin Powers movie. Snoezelen is a burgeoning practice, Lee tells me, but it makes one question our concepts of light and pattern and stimulation. She thinks one of many of the challenges with modern architecture, office design, room layouts, and other environments is that human beings are not being given the *right* kind of stimulation. It's not about over- or under- or bad or good, but finding your sweet spot. So if you want to paint a room orange or yellow—traditionally considered "bright" colors—you could play with how much white to add to the color, she says, to figure out the hue that soothes your senses.

I asked Lee about how mental health and well-being are being discussed within architecture and design as professional fields, because when I was in grad school studying public health, much of the focus was on topics such as how walkability in neighborhoods affects physical illnesses such as diabetes or obesity. "What about our inner lives, our emotional well-being and mental health?" I often wondered.

Lee's answer intrigued me. She has found in her research that architecture has inherited the same kind of bias toward behavior that the field of psychology inherited. In particular, a real bias from the colonialist mentality informs our spaces. When Europeans were setting up colonies in Asia and Africa, they were put off by the expressiveness of native peoples and the ways they exhibited joy—drumming and dancing and public merrymaking interwoven with everyday life. Europeans never wanted to show too much emotion or be too expressive or joyful or exuberant because it would blur the divide between colonizers and colonized. So European culture leaned in to restraint as a way of differentiating itself from other cultures that Europeans defined as savage. "That had implications for dress—and architecture," Lee says. "Architecture became about restraint, and early in the modernist movement people were dismissive of ornamentation."

For sensitive folks whose needs have been overlooked by the majority of society's structures—including design but also education, work culture, organizational practices, public health, and urban planning—it's important to carefully consider the "norms" we've come to compare ourselves to. If we feel out of sync with dominant preferences of color, design, relationships, and other factors that impact our everyday lives and sense of comfort and belonging in the world, it makes sense to consider that what has come to be defined as "normal" may not be in our best interest because of the umvelts we inhabit—and what is considered "normal" may not be in *anyone's* best interest, for that matter.

"The level of dysfunction in our environment is epic," Lee says. "There is a deeper problem in how we've designed spaces for relaxing and working. The problem is we simplify and isolate—we think windows will distract, but in fact concentration improves" when an environment has windows, she tells me. And the response of people experiencing Snoezelen therapy is telling, Lee says. "When you look at the elderly with dementia, they are losing their memory and their senses. So they're sort of losing their sensory tether on the world. The anecdotal evidence from Snoezelen suggests that people just 'wake up.' It's an underresearched area."

I heard as much from the Cooper Hewitt Smithsonian Design Museum's senior curator of contemporary design, Ellen Lupton, who was in the midst of a public sensory design exhibit when we spoke. The exhibition explored design that appeals to senses beyond the visual. Organizers were interested in things that use multiple senses and that play with the integration of senses across perceptions. For example, sound and vibration are closely linked because sound *is* a vibration. The show included experiences for people to enjoy and immerse themselves in while also being aware of their sensory faculties, as well as products that are designed to help people use different sensory experiences to solve a problem or improve their lives. For instance, there was a music player that translated sound into vibration that you can feel on your skin, and a "scent player" that created the scent of breakfast, lunch, or dinner.

Lupton is responsible for bringing *experiences* to the public, and she wants to help audiences understand, think about, and get excited about design. Her work has scholarly, practical, and deep components—thinking about design in all its facets and helping to educate the public on how design informs our world and how we inform design. "I'm interested in embodiment and embodied perception and the way even language refers to a kind of physicality of experience," Lupton says. "We're not just brains floating on a string." The world of design is very physical: everything we touch and use—a frying pan, a chair, or a book—has weight, sound, temperature. We're very intimate with a chair, a door, a car; a vehicle, for instance, is like a skin or womb, Lupton says. "I think there's a hunger for the real and for physical experience. I think people want to feel more."

Lupton also teaches graphic design, and she was struck by how young designers are intrigued by nonlinguistic components of experience. Something about the senses is drawing people in, waking people up to the full experience of being human, and intriguing their curiosities, she says. For instance, she has students from several different Asian countries who are particularly interested in engaging the senses. I noticed the same during the time I spent living in Asia and visiting Thailand, Korea, and Japan. There is a richness to public and private design—the streets, neighborhoods, vegetation, and apartments—that is stimulating in a soothing way. So when we think about our ideal environments, a broader consideration of the life of our senses must be deeply and intentionally integrated into how we shape and design those environments.

## Our Private Worlds

Feeling at home for me means being able to sink into a space beyond my own body, a kind of oneness with a room or place or even person. Joel Salinas, the Harvard neurologist with synesthesia, explains feeling similarly and has found that many synesthetes experience this as well. I suspect many sensitive folks across neurotypes can relate. Perhaps neurotypicals experience this when engaging in mindfulness meditation. For me, it is my absolute default state—almost like a state of awe—absent any anxiety triggers in that specific room or on that particular day.

I've experimented with many environments, routines, rituals, and human dynamics to gauge what my sweet spot of stimulation is, what makes my sensitive self feel at home and matched with my right umvelt amid a highly contrasting society. The goal is to find what enables your sense of "fit" and how you can sustain that feeling and compatibility. You know who you are and your particular sensitivity profile, and now it's a matter of tweaking a couple of factors so you can be your best, capable self, even if the larger world or reality differs from your unique umvelt.

For some people, the design and layout of their house, and their street and neighborhood, play an important role in regulating a sensitive nervous system and soothing an anxious disposition. For the moment I have found my particular sense of place living in peaceful, green, culturally diverse suburbs. I live in a quiet suburban complex surrounded by trees and hills, and my neighbors on all sides are from different continents. There is not much foot traffic, but I get to hear Guatemalan dance parties and smell Iranian cooking. I don't interact much with large numbers of people, but I sit at my desk and channel my default state of awe and stimulation into writing and advocacy. Aside from my beloved walks and time spent near the water, I don't need excessive stimulation; what pours forth from my mind and bodily energy is more than enough. My house is simple and comfortable, and I love it.

## Blue-Lit Seascapes and Starry Universes

Originally from South Korea, architect Kijeong Jeon came to the United States as a young man for his studies and stayed on to work in various locations around the San Francisco Bay Area. "When I came to Chico over ten years ago," he says, "I got a call [from a man] asking for interior design help. At first he just asked for help selecting a carpet. So I asked about the users, space, and activity—and he said this was a center for individuals with autism. I had never heard of autism before that." Jeon googled it and was shocked that so many people around the world were on the autism spectrum. What started out as a carpet request turned into a series of exciting projects thinking about design, well-being, and autistic clients.

"The sensory aspects—especially the lighting, acoustics, and tactile aspects—are significant," Jeon tells me. "What people on the spectrum are reacting to is very different from others." Jeon's Chico client did not have a strict timeline, and so Jeon spent about six months doing research on the task. He couldn't find any case studies in the United States but found several from England, as well as several therapeutic design examples from Germany and the Netherlands.

Jeon wanted to focus on creating a feeling of safety for the clients, through form, shape, and color. He asked himself, "What makes them feel secure?" In his research he came across the multisensory Snoezelen rooms, with their black lighting illuminating deep blue water tubes with bubbles and projections of outer space across the walls. They are designed to be soothing and energizing at the same time.

Color felt significant to Jeon, and so he focused on blues and magentas to create a feeling of safety and security for his autistic clients. Through his research on mental health and color, he discovered that the boring browns and beiges commonly used in hospitals would not be a good fit; violet shades are more helpful, so he went with that, much to his client's satisfaction. In one room, he installed violet and magenta lighting. Dubbed the "escape room," this was where folks could retreat when they were feeling sensory overload, a meltdown coming on, or like crying. Small and cozy, this room continues to be a favorite at the center.

Of course, it's important to note that everyone is different when Snoezelen or other therapeutic sensory design principles are being applied. So Jeon is hesitant to prescribe a set of rules or guidelines. For instance, I love blues and water and sea creatures, but you may like pink or orange and flowers and trees. Jeon does suggest, however, gradual transitions from indoor to outdoor so that the eyes can adjust, especially when implementing Snoezelen, since it requires the lights to be lowered. It's also important to be mindful of what kind of heating/cooling system is installed and how loud it may be.

Jeon is intent on understanding how well his designs work and has found that some of his large open, airy spaces are not a good fit for his autistic clients; smaller, tighter spaces are more effective and soothing environments for them. "I found it's very important to compartmentalize space," he says. He has found that dim lighting with a purple, pink, and fuchsia palette works well, along with well-lit decorative "bubble tubes," weighted blankets, and bean bag chairs.

Jeon also intuitively finds himself leaning away from pathologizing language; he doesn't use the word *patient* but rather *client*, because *patient* "makes it sound like they are sick," he tells me. Though in his native Korea he has found that such topics are still stigmatized, he sees his clients as merely "different"—not good or bad or normal or abnormal, simply diverse. Accommodating these differences has allowed unique and varied individuals to flourish and finally feel at peace in their homes.

## Intimacy, Relationships, and the Space to Be

### Denise and Tim: Filling in the Blanks for Each Other

"We've been friends since eighth grade and have been dating for real for six years," Denise tells me over Skype, with her boyfriend Tim sitting beside her. Denise is the medical student we met in Chapter 3 who was diagnosed as being autistic at age twenty-eight, years after graduating from a top East Coast university. I wanted to hear more about her life with her partner. "It's really convenient that we like being in similar environments," she tells me. "I actually do better in crowds and overstimulating environments than he does."

The pair originally hail from the South and grew up on the same street. Tim is tall, barely fitting into the Skype view. Unlike Denise, his southern accent dominates. He says he never thought of Denise as autistic until recently. "Because we've known each other for so long, it's never really occurred to me that it was a thing for her until she started studying this kind of thing. If we met now as adults, I would maybe pick it up because I feel as you get older you get more rigid and less forgiving for how people are. I wouldn't have said that she has $X$, $Y$, and $Z$. I don't really look at people in that light. That's just how people are," Tim says. "As a unit we balance each other out. I fill blanks for her and she fills blanks for me."

Tim has ADHD and says it's helpful for him—he knows how to navigate multiple kinds of environments beyond sitting at a desk. He reliably remembers the daily necessities for keeping the household in working order. He's social, too, and Denise feels she can rely on him as a steady sensory base. Although Denise says she can do long-term planning and order and schedule things in advance, she points out, "I can't do the dishes every day or remember to turn on the air conditioner."

Denise continues: "My big autism burnout hit when I was in New York, when we first started dating. I burned out at the end of the first year of med school, which is why I went for a diagnosis and why I needed accommodations." At the same time, Tim was dealing with an opioid addiction. "Our nervous systems were fried," Denise says. "We were forced to gain a greater understanding of how we show up in the world, in order to communicate and get through it together."

I ask them what tips they have for other couples. "Time apart," Tim says. He works in construction, wakes early, and is home by 3 p.m. When he gets home, Denise is often out studying. He thinks that's helpful and a sign of maturity. Denise says communication is important. "We're both really committed to authenticity. We both appreciate direct communication and we give each other permission for it," she says. "The way we communicate is much more direct and abrasive than a lot of people would enjoy. We're both on that same page. We're really good about vocalizing when the style of communication isn't working for us."

Denise says their biggest challenge is interfacing with other people. But the world the couple has created together has a reassuring safety and familiarity that others don't always quite understand. "People can't read us very well," Denise says. But Tim adds, "It feels like home."

I wouldn't say Denise and Tim are "typical" in their neuro-divergent dynamic and how they handle it—*because there is no typical*. What does intrigue me about them is how compatible they seem. They inhabit a shared culture—perhaps even a shared um-velt, given their sensitivities. I'm struck by a kind of sweet tender-ness between them. And it feels not too different from friendships I had in my youth at art school, where neurodivergence was no doubt common and kids weren't yet categorized.

### Isabel and Dan: Neurodivergent Family Roots

"I've always known I was neurodivergent," Isabel says. "And I probably chose Dan because I knew he was, too!" Dan smiles and says he didn't think about himself in terms of labels until "Isabel 'woke me.'"

Isabel, whom we also met in Chapter 3, is cheerful and laughing as she looks at Dan. She's the oldest of five and didn't realize she was autistic until her forties when her son was diagnosed. She studied biology and then worked as a forensic scientist in a police crime lab before becoming an artist.

The pair describe the "alternative" backgrounds they come from, growing up in a small coastal California town. Music and fashion were prominent in their lives when they met as teenagers on a plane to an exchange program in Europe more than thirty years ago. (Isabel says the first time she saw Dan at the airport, she was drawn to his sense of fashion.) They both describe themselves as being very much into aesthetics, and Dan tells me about Isabel's wide breadth of creativity.

Dan is a police officer, and his colleagues noticed that he is constantly humming, which Dan describes as his form of "stimming." Isabel stims, too, saying it helps her transition between tasks and activities. She identifies as ADHD and autistic. "My processing takes longer sometimes and much faster at other times," Isabel says. "Part of my unmasking has been letting myself off the hook a little bit and not beating myself up like I used to." She's been learning to treat herself in a nicer way, she says. "I'm more stochastic in my personality, and letting that out has been problematic!" But Isabel feels she's more creative as a result of letting parts of herself out. She finds herself in "flow" more often, for example, with her painting and writing. And tapping into community and finding connection with other neurodivergent people has been essential for her healing and for reducing stigma and pathologization.

To keep their relationship running smoothly, "checking in" has been important for Isabel and Dan, like notifying the other when one is running late. Isabel says she needs to be reminded to do such things; she's often late, sometimes spending four or five times longer than other people to get something done. Dan, on the other hand, is all about systems and routines and is always early. They laugh as they share this fact with me. But they cherish each other and allow each other to be who they are. Isabel "being late all the time annoys me," Dan admits. "And, you know, I find stickers from fruit all over the walls." But, he says, "We've learned to adapt for each other's differences. One of my biggest regrets is calling her lazy. That was before our real awakening. She's not lazy and never has been. But our awakening has brought us to understand and respect each other's neurological differences."

"I'm not the strongest verbal communicator," says Isabel. Sometimes she will try to verbalize, but what comes out is not what she means. "I tend to do better face-to-face." And sometimes she and Dan will write each other letters. They don't agree with the typical advice to never go to bed angry. "It's taken our whole marriage to get to this point," she adds.

Isabel feels like they used to be caught up in how society says marriage and parenting should be, but it's easier now. "Having kids has been really interesting," she says. "If you were to compare us to the average family, we're a bit different in the way that we parent. We're more hands-off, and it seems like other parents are dictators. Your child has their own passions and dreams, and we support our kids in what they want to do." Dan agrees, pointing out that they are attentive parents and always trying to be role models. "You do want to allow your children to find themselves and find their own ways of solving problems and their own ways of coming to realizations of who they are and what they want to be," he says. "*You have to integrate*," he concludes.

Again, like with Denise and Tim, I'm struck by how compatible Isabel and Dan seem. A joyful energy emanates as they answer my questions, and I can feel an embodied sense of how important their coastal culture of origin is to them. I again get a sense of a shared umvelt, a sense of home and place that is at once deeply nourishing and also stimulating, pushing the growth of ideas and stretching outward to innovate unencumbered by restrictive neurotypical barriers. And I agree that integration, as Dan says, is possible and can be accomplished in a healthy way with boundaries and neurodivergent identity intact.

## Couples Counseling for the Neurodivergent

"I was trained at the Asperger's Association of New England," Eva Mendes, a Massachusetts-based therapist, says. "After the DSM-V came out, they changed their name to Asperger/Autism Network. I trained there for a few years and then branched out on my own. I was looking to work with the adult autistic population and couldn't find anyone doing that kind of work."

Asperger's runs in Mendes's extended family. "Growing up, I wondered about certain things I would see in the family, and so when I went to AANE it felt like home. Autism is a neurological difference. I don't see it as a disease or syndrome or disorder. Each person is so unique, and I've never seen two alike. The work is never boring, and I always feel like I'm learning from my clients. They keep me on my toes."

A frequent writer on gender, autism, and relationships,

Mendes feels like she has what she calls a "shorthand" with the autistic population because of her family. She's able to understand both neurodivergent and neurotypical people, which is especially useful for the couples she sees. Her main focus in her training was on couples work; she was less interested in working with children to begin with, so she fills a unique niche within the neurodiversity field, given that adults are an underserved population in psychotherapy.

I ask Mendes how she helps her clients. What's different about a neurodiversity-focused therapy setting? What does she do differently? "It's framing the autistic person's perspective," she says. She tells me about a couple where the man is autistic and the woman is not. He often gets angry and resentful at his partner about what he views as broken promises. Mendes takes time to paint the bigger picture for the couple so they better understand what's happening. With this particular couple, the woman feels "emotionally starved," in Mendes's words. Actions and behaviors have accumulated on both sides, leading to lots of hurt—much like with any couple, neurodivergent or not. "He thinks maybe he should get a new partner," Mendes says, but she tells him that he'll run into the same issues, as she's seen several people in the same situation who are now on their third or fourth marriage.

The emotional reciprocity is a challenge, she says, as well as keeping in mind the context of particular behaviors. I'm nodding my head while she talks, as I'm familiar with it all. I can go into extreme anger in my head about what looks like a small incident, but to me it feels like a grave injustice. I'm narrowly focused on particular details of the event and I struggle to hold the larger picture in my mind. I had no idea that other people grappled with this. Talking to Mendes is unburdening several beliefs from my mind.

Mendes sees this scenario often: a person or couple will visit a therapist and be labeled "high-maintenance," especially if the autism is unrecognized. Or she will get emails from people challenging the possibility that they are autistic (which I also understand, because this is *our norm*). Others walk into her office unbelievably excited and liberated, glad to finally have a name for their experience. The quality of relationships can quickly improve. But the longer a couple doesn't know about the neurodivergence, the more baggage and lack of trust, Mendes says.

I ask Mendes about women in particular. "Their cases are sometimes more complex," she says, "because they so often have been misdiagnosed as borderline personality disorder or other labels." But maybe all the labels fit, she considers. "Maybe there's a dual diagnosis of autism and ADHD," for example.

Depression and anxiety are also clear overlaps. Mendes feels that depression, anxiety, and ADHD "fall under the autistic umbrella and traits. It's very rare to meet an autistic person who isn't also anxious," she says. Being able to pick out details and focus narrowly are potential gifts, Mendes says, but such things can also lead to perfectionism, which then can lead to depression. Focusing on what's wrong or "misaligned" all the time can be very anxiety-producing as well.

I call such states "flare-ups" and remind myself that my brain works like a camera lens—I can zoom in and out at will, and when I zoom out, I often feel immediate relief. Zooming in is my default—perhaps what makes me a sharp observer of other human beings—and so I use that ability in my writing and work. It's not as helpful in my relationships, where I can get very focused on small details that feel incredibly meaningful if I don't zoom out.

Mendes tells me about a young woman who is so extremely

articulate and self-aware about her own psychology that she could easily be the psychologist in the room. But she struggles with interactions in friendships and family. Mendes thinks that other therapists who rely on stereotypes—such as how autistic people dress—may not recognize this woman's autism. "Imagine, she's fashionable!" Mendes says, taking a jab at such unhelpful dominant narratives.

But Mendes believes her clients and their experiences, and this belief—this acceptance—holds tremendous power for healing. Even as she says this to me, I feel my body relax. I think that we sensitive neurodivergent women who can easily pass, or mask, may sometimes do so well at camouflaging our reality that we don't realize how tiring and taxing it is. Hearing Mendes say "I believe them" gives my whole body a sense of ease, like I don't have to fight to be understood and I don't have to hide to be accepted.

"From the beginning of the days of psychology, when it was very male-dominated," Mendes says, "there was this dynamic within therapeutic settings of the therapist being the expert and the individual being the patient." Mendes points out the hierarchical nature of older models of psychotherapy, but she sees individuals being their own experts. "It's about being humanistic and treating the person the way *you* want to be treated."

On the topic of trauma, Mendes reiterates that autistic women in particular can be more susceptible, sometimes because they are often more trusting and therefore have a harder time detecting the motives of others. Autism and post-traumatic stress disorder can coexist, and sometimes that complicates a diagnosis, but Mendes sees them as co-occurring. When she's working with younger adults, she often invites the parents to be involved in the therapy. "They've been raising them," she says, so she wants to get an overall picture. She also talks to anyone else who is involved, such as a psychiatrist or an executive functioning coach. Sometimes people are taking medication for co-occurring ADHD, depression, or anxiety, but she reiterates the importance also of "just listening."

Mendes and I talk about meltdowns or "explosive behavior." "I work with clients on reframing what happens," she says. She educates her clients about "energy quotients," that is, thinking through how many "energy points" a person may start off with in the morning and then paying attention to which activities or tasks drain that energy bank. Even as she explains this, I start realizing for myself that simple things in my day may drain a good chunk of my available points. "Then you can replenish by doing something you enjoy," she says.

The challenge is that many people don't recognize the triggers and stressors as they happen, and so they pile up and can lead to a meltdown. Lots of people have friends they can call and process with, but many autistic people don't call up a friend to just talk, Mendes says. A therapist may be the only person who knows what's going on; for instance, even a neurotypical spouse may not be aware of the triggers that are affecting a neurodivergent spouse. "But there's a reason for the meltdown," Mendes says. "It's not coming out of nowhere."

Mendes integrates this awareness of trigger buildup into her counseling sessions; she is careful, for instance, to always turn off lights for her clients. Outside of the office, she may encourage them to wear sunglasses when they are running errands, and she reiterates what many occupational therapists told me—exercise is crucial. But key for clients is recognizing when the triggers are beginning to build. Being mindful of who they spend time with is important, as is paying attention to boundaries, she says. Some of her autistic clients have tended to tolerate friends who are unkind or talk endlessly about personal problems or negative topics, and the autistic client will think they should "take it" as a way to be a good friend. Mendes reminds them of the importance of self-care and boundaries in such situations.

In many ways, what gets magnified in the homes, relationships, and intimate daily lives of neurodivergent people all over the globe is indicative of the many ways that society could learn from them and make changes that would benefit everyone. Better design, sensitive aesthetics, relationships and families that celebrate difference, and therapists who don't seek to pathologize every departure from the norm—these are all great ideas. I look forward to the day when what is currently a hidden sensory world for many people becomes the global norm. What are labeled as sensory ailments actually hold promise for healing a fractured and traumatized world that is in desperate need of repair.

## Sensory Tips for Home

Consider how your choices regarding your sensory lifestyle may or may not be aligned. Here are some tips:

- Identify your ideal color palette. For me, salmon, turquoise, black, white, and gray make me feel at ease, so I prefer them not only in my clothes but also in my work, branding, logo, and home decor. I feel soothed by the colors but also the consistency with which they appear in my life.

- Include nature in the mix. If you don't walk or do other exercise outdoors regularly, start now. If you're not able to see trees from where you live, consider adding plants or other green decorations in your home, such as a "living wall" of greenery.
- Be up-front with your partner, or potential partner, about your sensory needs—such as preferences about lighting, the sound of the bathroom fan, or the amount and loudness of music in the house. This will prevent drama later on. Up-front communication is key to relationship harmony.
- On that note, question the lighting, color, and neighborhood decisions in your life. Some of your happiness and mental health could be severely affected by where your house is located on a street—is it on a noisy corner, or tucked quietly behind a front lawn? Is it next to a bus stop? Are the lights in your garage too bright? Is the outside paint too bright, or not bright enough?
- Seek out a therapist who understands sensory needs—this applies to individual therapists and couples counselors alike. Too many folks suffer for far too long and endure needless conversations and interrogation that simply miss the point. With this book as your armor, you don't have to wait for the entire international scientific community to update their research logs. Explain your sensory needs, see whether your therapist or doctor gets it, and proceed with confidence.

## Chapter 7

# Work

The question of work—or making a living—is at the heart of so many struggles for neurodivergent folks. We think differently and process differently, so we work differently. It's a challenge to time our abilities with a clock when overwhelm or underwhelm, boredom or overexcitement are rotating realities. For myself, I get creative whims early in the morning and sometimes late in the evening, but I also need to get my family ready for school in the mornings and arrive at meetings when expected—all of which confuses my sensitive self. But at the same time, without that sense of being plugged in to the world, we are also at risk of isolating ourselves and becoming lonely. It's a challenge to strike a balance, and many of us become therapists, writers, and entrepreneurs to help alleviate some of that struggle.

*The subject of work is a tender one for me. In my midtwenties, I was fired from one of my first journalism jobs. The pace was so intense and demanding, I wasn't the only one to leave, but it still hurt. Then I became a freelance journalist, which worked quite well for me, since I had the flexibility and freedom to pursue different topics. But when I attempted to move up the ladder in my early thirties and landed a job as a senior editor—albeit at a very dysfunctional early startup—things came crashing down. My sense of overwhelm and brain fog was like nothing I had ever experienced, and I was again fired after just six months. A year later I took a low-paying job in a more arts-oriented organization—thinking that was where I would find "my people"—but instead I was burdened with administrative tasks. Again, I was fired after a month. The hardest part throughout the trials of these various jobs was the overwhelming sense of loss, confusion, loneliness, and uselessness I felt. I had zero vocabulary about neurodiversity—and no one else around me had that vocabulary either. I felt judged, criticized, and undervalued by my coworkers and by society as a whole.*

*When I reached a state of delirious confusion and loneliness about what the hell was going on with me, I finally started opening up to others. Starting with my immediate family, I told them about the news reports of women—who sounded like me—who had sensory processing differences. Compelled by what I was reading, I created a Facebook event page inviting people to get together and talk about neurodiversity. I held the first Neurodiversity Project gathering in a small Berkeley studio filled with people craving honest discussion about how our minds seemed to work differently. Leaders from corporations and academia opened up alongside artists, singers, and activists. A buzz grew within me that I hadn't felt in years, and it grew stronger with each gathering—and that was it: finally, at last, my neurodivergent working style as a writer and entrepreneur was flowing with something that people wanted and needed and where I could run a thriving venture.*

## Opening Up the Conversation

In my 2017 feature article in *Fast Company* "What Neurodiversity Is and Why Companies Should Embrace It," I put forth suggestions for how to make the workplace more neurodiversity-friendly. That article led to dozens of inquiries, tweets, and emails, especially from women. At around the same time, *Harvard Business Review*, *Forbes*, and other publications were starting to catch on to the dire need of addressing these topics. At the center of my article was a woman named Margaux Joffe, head of production for Yahoo's Global Marketing Department at the time, who also launched the company's first neurodiversity Employee Resource Group. Joffe has ADHD and previously founded the Kaleidoscope Society, a platform for women with ADHD to share and celebrate their successes in life—think artists, scientists, human resources folks, designers, and a whole range of career positions.

"My advice to neurodivergent employees is to learn as much as you can about how your mind works in order to design your daily life accordingly and be able to effectively communicate what you need at work and at home," Joffe tells me. "I lived with undiagnosed ADHD for twenty-nine years, so the diagnosis alone has helped me tremendously in my career. Simply understanding how my mind works differently, I've been able to let go of how I thought I should do things and accept myself for who I am."

Joffe explains that what often stands in companies' way is

a lack of knowledge for how to approach subjects such as neuro-divergence. The growing "neurodiversity movement," focusing on *diversity of mind*, just as activists in the 1960s and 1970s fought for racial equality and gay rights, is spurring an increasing aware-ness that is likely to put much more pressure on workplaces to understand neurodiversity, adhere to policies, and respect neuro-divergent folks—especially women and people of color, Joffe points out, "who already feel like they have to work harder to overcome unconscious bias."

It took Joffe a year to "come out" to her boss about having ADHD, because, she says, "I wanted to be able to prove my-self free of any additional bias." But once she did, and told him she wanted to launch an Employee Resource Group for neuro-diversity, "he supported it 100 percent." She continues, "Many times, the only thing holding us back is thinking we need to work like others. Build on your strengths and be fearless. This goes for everyone."

For myself, as a sensitive neurodivergent woman, my primary concerns relate to social norms around extreme extroversion, expectations of high-speed productivity, overstimulation, and overemphasis on factory-style executive functioning. My ideal setting is a quiet, natural light–filled place surrounded by green-ery, with easy access to colleagues when needed. I value mentor-ship and learn well one-on-one. Being myself often looks like lightning bolts of inspiration, needing to pause a conversation to write down an idea, emailing or calling someone spontaneously, or needing to take a quick rest when I'm overstimulated or work from home when I have a headache. Susan Cain has made some of this language much more accessible and acceptable through her Quiet Revolution website and Quiet Workplaces and Quiet Schools programs. Walmart's most productive distribution center is staffed almost entirely by neurodivergent employees. SAP has an entire Autism at Work program, and Microsoft now holds annual Autism at Work events. Things are starting to shift, but more is needed.

## Work and Temperament Rights

Our internal makeups matter. *A lot.* How I spend my days thinking, reflecting, and responding to the world correlates to my mental health, the mental health of my family, and the overall health of my workplace. Each of us creates our own reality, largely dictated by our internal perceptions and how those perceptions are shaped by people in power. As Elaine Aron frequently likes to bring up, if we as sensitive people are raised in supportive environments, our unique gifts are often fostered and allowed to thrive; but if we grow up in negative environments, we often develop depression and anxiety. And so the interplay between inner and outer is significant and largely holds the key to the overall health and well-being of us as individuals and of our wider communities.

The idea of "temperament rights" brings into the equation a consideration of our inner constitutions in every sphere of life—work, family, school, education, sports, religion, and more. The unique individual makeup of each person deserves its own articulation, respect, and corresponding accommodation. Note that this is not the same as every person getting exactly what they want all the time. But it does mean, for example, that when a person starts a new job, it is carefully noted in their personnel file whether they identify as neurodivergent, and if so, which neurodivergence, as well as their workplace preferences and needs. That person's particular sensitivities need to be acknowledged. They will then know that their internal reality is acknowledged at work and they can thus resort to the language of temperament rights and neurodiversity to sort out any issues as they arise, as well as advocate for themselves.

When I got back in touch with Joffe a few months after our initial interview, Yahoo was in the midst of its merger with AOL after both companies had been acquired by Verizon. It was a tough time—and in many ways traumatic. People were being laid off, and change was everywhere—from management structure to the redesign of office space. It was a significant time to talk about neurodiversity at the company, as unexpected shifts often allow for new ways of leading and operating, and Joffe's team advocates for neurodivergent employees in a number of ways. "We've been bubbling up some questions and concerns to leadership—everything from office moves and layouts to accommodations," she told me. Her team was making recommendations to the team in charge of real estate and workplace, such as designating quiet spaces, working with individuals who need accommodations, and making noise-canceling headphones available.

"I feel like we're just beginning to see the start of a new wave of how mental health and neurodiversity are going to be looked at in the workplace," Joffe says, "because traditionally the workplace hasn't always been best-suited for anyone who doesn't fit the mold, so to speak, but I do think things are beginning to change. Once we launched the ERG [Employee Resource Group] it opened up a Pandora's box and people came out of the woodwork, sharing their stories. It was definitely the case that people were dealing with things on their own and keeping to themselves." Launching the neurodiversity ERG was a big, brave, and vulnerable move for Joffe. It took courage to put herself out there, and ultimately that courage led to her being on stage in front of eighty-five hundred people to share her story.

Indeed, sharing stories is often the first step for helping changes take hold in a workplace. After Joffe formed her ERG and started speaking openly about the gifts and challenges of her neurodivergence, dozens of others started speaking up as well. "In our department alone, through launching this group," she says, "I found out a few of my colleagues are also neurodivergent—and I would have never guessed! I would've never known what they were going through. It's been so powerful because people from senior leadership are also coming forward as having ADHD. I don't know if there are any other companies out there specifically with neurodiversity ERGs. I know there are autism working groups, but I'm not sure about neurodiversity more broadly. I've seen disability and health and wellness, but not neurodiversity. It's an open frontier—a lot to learn and a lot to do."

Embracing neurodiversity at work also allows for a narrative shift to happen, because colleagues find themselves surprised by the mismatch between *what they think a neurodivergence might look like and how it actually shows up in a person*. It is essential that all workplaces—companies, organizations, nonprofits, government offices—open up this "Pandora's box" so that the buried treasures and gifts that neurodivergent folks possess can emerge and help entire teams. Senior executives, managers, bosses, board members, and others have a significant role to play in creating work cultures and spaces that allow for the language of neurodiversity. This can have a profound ripple effect throughout companies and society at large. The opportunity to transform social norms is real—*and tangible.*

## Adobe: A Lesson in Sensory Design

Silka Miesnieks is another bright star in the neurodivergent underworld of tech and design in Silicon Valley. I saw her speak at a design conference in the Japanese innovation space Digital Garage in San Francisco. The conference was about artificial intelligence and the intersection with design, empathy, emotions, and human impact—and to my delighted shock, some of the first words she delivered were about her being dyslexic and about the senses of proprioception and interoception as they relate to design. I felt so lucky that I had decided to attend the conference that day, walked into that downtown San Francisco tech space, and happened upon someone so aligned with my research interests.

Miesnieks is now head of Emerging Design at Adobe. "I actually got diagnosed a couple years ago with ADHD, and my son has been diagnosed with that, too," she tells me over the phone a few days after the conference. "At first, we thought he was dyslexic, but then it turned out it was ADHD. And often those two go together. So it's been quite a journey having spent most of my life not knowing I've got ADHD and then not getting help with dyslexia because I wasn't quite 'bad enough.'"

Miesnieks learned early on that she had a gift for spatial ability, and from a young age she began making 3D creations. "I ended up being an artist because that was the best way I could communicate—through visual imagery." She studied art, which led her to technology and animation and then graduate school in industrial design. "It's all around these connections between how we feel and the spaces that we live in. And I feel very attuned to that." In the past she led startups as well as teams in large companies. She's always been more of a leader than a follower, she tells me, attributing that to the ADHD.

I ask Miesnieks about how she went from identifying her own neurodivergent inclinations—seeing things differently, thinking about design differently—to action. How does one go from identification to input and effect? First off, she says, "I think having been an entrepreneur and having to get comfortable with being uncomfortable and always being insatiably curious helped me overcome fears of failure—having an inkling and just going with it." Then, to arrive at what she and many others in the field are calling "sensory design," she went from discussing some ideas with a designer friend of hers to asking experts in various fields some questions about this new direction in her work. "I constantly test my assumptions," Miesnieks says. "I don't feel like I have all the knowledge and expertise that is needed." So she reached out to leaders in the industry, and "everyone said this is something that we need to be doing."

Psychology has focused on the mind, but there are newer insights now around the body, movement, and kinesthetic styles of learning, so she is seeing a shift. "This is a nice time, where the technology can reflect society's mood. Our cultural shift has also been to think about the whole person. And the way my kid and I like to learn—and the more research I did—I saw how we are not edge cases. So even things like voice-to-text applies to so many more than just dyslexics."

Miesnieks was able to follow her gifts and inclinations, become an entrepreneur, and now also channel her knowledge and style into a large company environment. "It's been a really big thing for me," she says. "My first reaction when I found out about the ADHD was anger. 'Why wasn't I diagnosed in college? That would've helped me. I could be so much farther ahead in my career. I could have had an easier time.' And then I got over that and I started to see the benefits. *I was never able to recognize them before, because I didn't know I was different. I thought I just had to try to be normal.* The ability to connect lots of different abstract ideas together and come up with concrete solutions and combining many different thoughts and concepts and seeing them through all different angles all at once and then connecting them together—I didn't realize that was unique."

Miesnieks also points out the importance of who you are surrounded by. Her first boss at Adobe, for example, was very encouraging. "He recognized that I got things that no one else got," she tells me. Rei Inamoto, a New York–based advertising and innovation executive who hosted the design intelligence conference that day, also told me as much when we spoke a few days later. Recognizing the high proportion of entrepreneurs with ADHD in Silicon Valley, he prompted me to think about the ecosystem surrounding the people who succeed in that environment.

Miesnieks's second boss is the head of all of Adobe Design. "All she does is encourage me to be me, which is all I needed. Just trust," Miesnieks says. "And she has to have enough security in herself to be able to do that, and thank goodness she does, to be able to say, 'I'm just trusting you.' And I've grown because of that, and I'm pretty much given my own freedom. It's unusual, really unusual."

I ask Miesnieks whether she's starting to see more conversations about embracing mental differences at work, especially in Silicon Valley or the global design and tech startup ecosystem. "I still think ADHD has a bad name," she says. "I hesitate saying it. I can now say I'm dyslexic and not feel bad; I feel that stigma is gone. But the stigma with ADHD is still there. It's not well known, and it's only because I have it myself that I realize how different it is for different people." She reiterates how there's a prevailing outdated stereotype that ADHD means being hyperactive and always bouncing around. "That needs to change," she tells me. "What I'd like is that ADHD become associated with good things and with abilities that are valued in the company and in the community as a whole." She's found colleagues at Adobe who are also ADHD, and they compare notes and talk openly with each other. "I think when it's an open conversation and when there's a support group, you realize an amazing amount of people have it."

When I ask her about women in particular, Miesnieks says she sees immense value in women with ADHD being able to talk about it with each other and connect. She's part of several women's groups and says ADHD stigma is one of many issues women face in the workplace. She notes that "women are very good at supporting each other." When she brings up ADHD in her women circles, she feels immense relief and finds that everyone has a story related to some kind of mental difference.

In the end, she says, "I'm just really grateful for being ADHD. It's given me such creative abilities. I want to put that out there for other women as well."

## Masking at Work: It All Starts
## with the Job Interview

The friction between our modern-day neurotypical expectations in the workplace and the reality of human neurodiversity shows up early—the job interview process. Previously, when I went into a formal setting such as a job interview, I would unconsciously go into masking mode—speaking and gesturing in ways that, in hindsight, didn't feel good or natural to me. Like many who try to make a good first impression, I would present in a way that I knew was expected of me, but I would feel exhausted and wiped out the second I got home. I was essentially putting on a neuro-typical mask but didn't realize it—and that's not sustainable for the job applicant or for the boss or company as a whole.

For me, I find myself having to split between what I'm perceiving on the surface of a work-related conversation and what my mind and body are picking up under the surface. Take this hypothetical job interview:

**GREG (JENARA'S POTENTIAL BOSS):** What was it like living in Nepal?

**JENARA'S INSIDES:** Is he asking whether I'm unreliable because I've traveled so much? Is he trying to assess whether I'm going to pick up and go travel again? I see his balding head and heavy eyes and wedding ring—is everything okay at home with his family? I feel really sad. What happened to him? What were his parents like?

**JENARA'S OUTSIDE:** It was awesome!

**GREG:** How do you approach interpersonal challenges at work?

**JENARA'S INSIDES:** Oh God, does he want the whole story? My extensive take based on the latest research and writing and my totally alternative, unconventional thoughts on the modern workplace? Will that be too much information for him? Wait, I'm really smart— maybe I should just share all of that? But wait, will that make me look like I can't follow neurotypical norms and hence am somewhat of a rebel who can't follow marching orders?

**JENARA'S OUTSIDE:** I'm a strong believer in open communication, being clear from the start, and having dialogue when things don't go as expected.

Does this sound familiar? As neurodivergent individuals, we are almost constantly having to hold two realities simultaneously, because we pick up on so much external stimuli and have the repeated unfortunate experience of having our ways of interacting shunned and rejected.

## Mental Health at Work: Leading the Conversation

I'd like to see neurodivergent entrepreneurs, designers, researchers, media executives, and policymakers inserting more of their full selves into the world, and I'd like the rest of us to celebrate those full selves more. It's also important that journalists, conference organizers, and university departments cover stories about neurodivergent folks and invite those folks to the stage.

Leigh Stringer, for example, is a writer, designer, and conference organizer who is shedding light on how office design affects well-being. She has focused on the impact of artificial lighting on mood and sleep, as well as how nature and biodiversity reduce anxiety and depression. For example, "living green walls" covered with a variety of plants evoke positive psychological responses, and hearing multiple birds sing has been found to be more relaxing than hearing one single bird. Stringer also says an important new area of research "has to do with choice about how, when, and where people work. Choice and autonomy can go a long way in improving health and performance," and she has collaborated with the Harvard School of Public Health and elsewhere on new areas of research on the intersection of work, design, and well-being—with applications that can benefit neurodivergents as well as help determine how neurodivergents can have more input.

Over in the UK, Barbara Harvey at Accenture has helped spearhead a deeper look into mental health at work and how workplaces can better support people and remove stigma. "Over the last five years we've seen a tremendous increase in the public debate around mental health," she writes to me. Her team ran a number of surveys and found that increased media coverage and an awareness campaign run by the royal family and partners have helped people feel more comfortable talking about mental health now as compared with a few years ago. "The fact that people feel they can speak more openly, combined with a better understanding of how mental health touches people's lives, has created a force that is requiring government and employers to focus on solutions and approaches." I ask her about how such change happens and how stigma begins to fade. "By starting a conversation," she says. "When leaders and colleagues speak openly about the topic, they help make it safe for others to do so." Princes William and Harry, in particular, have shared openly about challenging times in their lives, how they sought help, and how talking with colleagues helped them feel less alone and more connected.

Harvey says programs need three elements: buy-in and support from a leadership team, resources in place to meet the needs of employees, and awareness-building through everyday conversations about mental health. In the UK, 15 percent of the Accenture workforce, about eighteen hundred employees, has been trained as "mental health allies," that is, "colleagues others can approach in the knowledge that their discussion will be kept completely confidential." Accenture's goal is to have 20 percent trained as allies and 80 percent trained in basic mental health awareness.

"For a long time mental health at work has been seen as a minority issue," Harvey writes. "Many companies have approached mental health in a reactive way, responding at the point when someone needs help." But, she says, reframing mental health as something we all possess and recognizing that it can fluctuate among and within individuals help to normalize the conversation, and, ultimately, she wants to see companies be more proactive. She especially has faith in millennials as a demographic that prioritizes mental health.

## The Writer's Path

"I'm thirty-six and have been working as a freelance writer for half of my life. My grandparents live in midtown Toronto, and I'm a very autistic creature of habit, so I came back here to live," writer Sarah Kurchak tells me over the phone. We found each other on Twitter—Sarah wrote a powerful and tender article about her relationship with her mom and the kind of unconditional support she received throughout her childhood. When we spoke, Kurchak had just returned from her annual adventure of covering the Toronto International Film Festival (TIFF). "I have weird intense areas of focus that are very rewarding," she tells me, also referring to her career as a music journalist.

"I knew I would never be able to do the work I wanted to do if I had to pay off student loans, so I made a naive gamble to intern at television stations and magazines and trade my labor for on-the-job education, and I ended up at a music magazine. I don't have any formal training, but I picked up a lot of skills," Kurchak says. Her story is somewhat common across neurodivergent-identifying folks because of her ability to focus intensely and her insistence to learn differently from neurotypicals. "Journalism was always an accident. Recently I've been shifting to personal essays and critiques, because I think that's where my strengths are."

Referring to her experience at TIFF, Kurchak explains that navigating the film world has surprising benefits and a compatibility with her neurological makeup. "At a certain level, the film industry is so nakedly upward-grasping that everyone's motives in social conversations are so obvious that they're easy for autistic people to track. And it's actually kind of more comfortable, because you know what everyone wants from you. And you know when they don't want it anymore. And so it's kind of a relief. Everyone else is stuck on 'Oh, that's so fake.' But lots of social interactions are fake. This is actually sort of weirdly honest in its own way, and I can handle that." I know what she means. When I'm at conferences and everyone is networking, and I *know* it's a sport, that can be empowering.

But there are other components Kurchak can't handle, she says. "The crowds are getting to me more year after year. I think last year was the first year I almost melted down in line for a press film." She also mentions glaring lights and noise, with loud screens and more. "I'm also starting to feel a little isolated by all the cliques. I feel a little lost when I see everyone else connecting and networking and I'm just lonely." She wishes all the neurodivergent folks could come together at the festival for a shared purpose such as talking about how to navigate the sensory overload.

Kurchak was twenty-seven when she was diagnosed. "We always knew I was different, but just never had the language for it. I can't remember the first time I heard the word 'autistic.' Doctors were analyzing my gait, I felt awkward, but sometimes too smart, and my parents were just practical and did whatever seemed to work for me as a person. I credit everything I've done right to them. I was lucky enough to be born to them."

In her early twenties she started to hear about Asperger's, and one day while reading a checklist of "symptoms," she resonated with every single item. She started to look into testing, and nothing was available for people over the age of eighteen. She felt like the Canadian government health system couldn't help, and she couldn't afford private testing. For a while she accepted the fact and moved on. But when she had a meltdown over something relatively small, her mom rushed her to get tested, telling her, "I know we can't afford it, but this is something you need to know because you're not going to be able to piece together anything else if you don't have the foundation of understanding of what's actually going on with you."

After a lifetime of knowing that "something was up," Kurchak and her mom were able to finally understand every detail during the diagnostic process. Kurchak felt that diagnosis was important because she wanted to feel differently about her childhood, that is, she wanted to feel proud instead of beating herself up for her failures, as she puts it. Diagnosis finally helped her understand why she felt different both as a child and as an adult in work settings.

"I still don't feel like a grown-up," she says of her life as a freelance writer in Toronto, living with her husband in a small apartment. But she's found a sense of community with other music journalists. "These people are so much weirder than me that I can totally fly under the radar."

## Burning Woman

I first encountered Lucy Pearce, Irish author of *Burning Woman* (2016), on Facebook after reading that book. She eventually was diagnosed with Asperger's, and with time she came out about it. "I was always aware that I didn't fit in," she tells me in an email exchange.

> I never really belonged, despite putting so much effort into being "normal." Though "normal" was my goal, I never seemed to be very good at passing—and it took all my energy but very little of my interest. I wanted to fit in socially in order to be safe, nothing more. I have always really enjoyed being by myself—reading, thinking, drawing, writing. I connect deeply to a few special people, but socializing just isn't what I do for fun.
>
> . . . I can now see myself identifying my "not normalness" through my writing in the early part of my career, especially in my book *Moods of Motherhood*. But I had no

name for it. My mother came across Elaine Aron's work when I was a teen, so I knew about HSPs before I had my children. I knew this term was the closest to identifying my issues and I used it in my book, *The Rainbow Way,* as a way of describing my struggles with devoting myself to motherhood, and my deep-seated need to create. HSP is a phrase, I think, that united a lot of us neurodivergent folk before the awareness of Asperger's in women started to appear. This high sensitivity impacted my whole life. But I was always a good girl, because I was so scared of causing conflict or upset that I couldn't handle, and was always top of my class, so I never rang any bells for teachers or doctors. I managed to keep my struggles reasonably private and well-hidden, only interacting with the wider world when I was able—I was the queen of masking. Or at least I was until I was trying to manage mothering three children and running a creative career.

Pearce's writing and art are poetic and mystical, infused with themes of spirit and both oppression and liberation. Her work is also a call for women to unite as they embrace their longings, sensitivities, artistic natures, and more. I found her work particularly compelling and helpful in the beginning of my neurodivergent awakening, even though she wasn't using those terms and she hadn't yet learned of her own neurodivergence or the larger neurodiversity movement. She knew she was sensitive and also identified all three of her children as sensitive.

But the sensitivity and struggles with one of our girls were much greater than either my husband or I could relate to or manage. She is extremely intelligent, hit all her milestones—so doctors and school just put it down to shyness and lack of firmness in us as parents. But I knew it was more than this. So I started researching, looking for answers. I had done a huge amount of reading on her behalf—Aspie [Asperger's] heaven!—but hadn't considered it for myself. After all I had been to university, I was married with kids, I was an established writer . . . and I was in my mid thirties. Surely something like that would have been picked up years ago. I, like so many people, had only had images of Rainman [the movie in which Dustin Hoffman plays an autistic savant] . . . as my touchstone for what autism looked like. But then two writers and entrepreneurs [in my field] came out as being Aspie. I identified so strongly with them and their work . . . and struggles. If they were Aspies and were also able to visibly function much of the time . . . then perhaps I might be too.

Much like Sarah Kurchak, Pearce had to eventually seek out private testing for herself and her daughter as public services were lacking. "The clarity and certainty it [diagnosis] has brought me has been immense. The relief. The knowing I no longer have to try—and fail—to be normal. That I am free to live life as me. That I am not broken or a failure. Nor am I imagining things. I feel like I am at last at home in myself, after 38 years. But I am still adjusting and processing the whole of my past through this new lens, and it takes time."

Pearce continues, "The pieces of the puzzle of me have finally fallen into place and I have permission to be myself in my own way at last. This has helped me immensely on so many levels—finding the right medication, a wonderful therapist, a peer-support group, and being able to explain my struggles to others."

Pearce also published a book that detailed some of her journey called *Medicine Woman: Reclaiming the Soul of Healing* (2018). In it she pays particular attention to physical ailments, such as fibromyalgia and chronic fatigue syndrome, that sometimes accompany sensitivities and neurodivergences—and that are often dismissed by the medical system because of gender bias. She is also the founder of Womancraft Publishing House, where she helps other women share their gifts through art and writing.

My "outsiderness" is actually an advantage; I am already neurologically seeing our culture from the standpoint of an outsider, whilst being aware of it from an insider perspective as a human, so am able to articulate this unique perspective. Once I was accused of being like Dianne Fossey observing gorillas in the way I was observing the dynamics in a group before I felt safe to share. This was said in a cruel accusatory way, but actually it was true. I do study humans and our culture, all the time. I am a thoughtful detached observer. I see the damage that our culture does to so many who don't fit in to the white, middle-class American dream of capitalism, I see the damage it does to our greater ecosystems. And so my utopian, pattern-seeking, systems-loving, philosophical-scientist, artist-dreamer self longs to help find other ways of living—healthier, gentler, richer more nourishing, nurturing and creative alternatives to this dull, grey, loud, violent culture. This is the driving passion of my life and work, and thanks to the internet I am able to connect with so many exciting, passionate, creative dreamers around the world who are doing the same.

She concludes with this thought: "A woman in one of the Autism groups I am a member of recently said that autism is not a disorder, it is a community. I like that and am living into the truth of it more each day."

## On Sensitivity at Work

To tie some of these themes together, I want to share something Margaux Joffe told me. She has since become the associate director of Accessibility and Inclusion at Verizon Media:

> I think we as neurodivergent people have higher levels of empathy—though technically it's called "emotional dysregulation." It's not necessarily that we feel more than neurotypicals, but the automatic mechanisms for regulating our emotions don't work the same way and so we feel higher highs and lower lows. It's not that neurotypical people don't have that capacity to feel, but they're more regulated and so they don't feel the same levels. So I think being empathetic people, we pick up on things that are unspoken and we pick up on the energies of those around us. Neurodivergent individuals are great people to go to when you want to get a temperature check on how people are feeling in a company, because they're probably picking up on everything. We're also more unfiltered and say what we think and feel and have a lower tolerance for bullshit.

Does any of this surprise you? I was thrilled to hear Joffe say this. Here I was talking with someone with major responsibilities at a leading corporation who was basically articulating where the many threads of neurodivergence actually *converge*—that is, in the area of *sensitivity*—and the case for embracing temperament rights and neurodiversity at work is born. (Not only that, but she wasn't afraid to infuse corporatespeak with phrases like "picking up on the energies of those around us" and "lower tolerance for bullshit"—*my kind of woman.*)

## On Sensitive Leadership

I also asked Susan Cain about how high sensitivity and well-being come up for her clients in workplaces or elsewhere, especially in leadership positions. First, she tells me about the widespread complaints about open office plans, which enjoyed some popularity with the rise of Silicon Valley but are now seeing backlash. Such plans are a serious concern for many HSPs and introverts, often provoking debilitating anxiety. "I can't emphasize enough how much it bothers people," Cain tells me. "There's a special level of emotion and urgency around it coming from the people I work with." Being constantly bombarded with sights and sounds and being observed and judged and on display—*it's really not working for people,* she reiterates.

We often focus on neurodivergent employees struggling in a cubicle environment, but the truth is that many people in senior executive positions are navigating high sensitivity as well. Cain has been working with leaders across fields on introversion, sensitivity, and the broader themes of her book *Quiet.* "For sensitive leaders who learn how to manage their sensitivity in the often sharp-elbow world of the workplace—once they've learned to manage it, we know the upside can be quite powerful," she says. "The ability to feel what's happening on your team and talk to one team member in one way and another in a different way— the sensitivity can be a superpower in that way."

What Cain says next is perhaps fundamental to the sensitive neurodivergent's quest to power through, plug in to the neurotypical world, thrive, and make a unique impact. "I think the tricky part is that it's hard to use that superpower when you're feeling overwhelmed or anxious. Feeling those feelings can make us behave in a self-absorbed or selfish way even if that's not who we really are. So what I see is that the first order of business is managing the tendency to feel overwhelmed or anxious. And then once you've got that nailed down, then your superpower is free to express itself."

And it is this superpower that neurodivergent people are now learning to harness all over the world. We have needed a common language, a common vocabulary, to take with us to work and beyond that enables all of us to engage with the world at a depth that so many of us have craved for far too long.

## Your Neurodiversity Footprint at Work

As I said before, I'm not a therapist. Nor am I a corporate executive. But being someone who has had to navigate work environments that essentially set up neurodivergent people to fail, I have some opinions on what companies could do better. So here are some of my practical suggestions:

- Make clear during all new hire orientations that neurodiversity is understood, welcome, embraced, and accommodated. This will help people feel comfortable being themselves and encourage them to do their best work.
- Train all staff in the language and framework of neurodiversity—to start, take one hour out of a workday to introduce the concept, give examples, and emphasize its importance, relevance, and significance.
- Consciously create opportunities, working groups, and job positions that rely primarily on creative thinking and less on linear thinking so that neurodivergent employees can excel and not waste time trying to catch up to neurotypical expectations.
- Create physical spaces—quiet, private, open, shared, etc.—that accommodate a range of work styles.
- Integrate the natural environment into the work environment as much as possible. Neurodivergent employees need more time to regulate their nervous systems, and so on-site gardens and patios, light-filled windows, and proximity to parks are crucial to ensuring that creative, out-of-the box ideas emerge. The last thing you want is a treasured employee burdened by unnecessary headaches or other discomfort due to work environment mismatch.
- Offer regular, ongoing small-group meetings on mental health, burnout, sleep, communication, boundaries, and organizational tips and tools. Workplaces hold the power to drastically reduce stigma, and that change reverberates outward into families, communities, and society at large.

# Conclusion

*Now is the time to build human ecosystems that are sustainable for neurodivergents worldwide. Countless individuals, projects, organizations, and movements are working to change the landscape of opportunity for people like us. Women and men are reinventing health care—specifically mental health care—and a range of scientists, researchers, artists, and community innovators are leading the charge. From the origins of gender bias in the evolution of psychiatry to the reframing of diagnoses in the DSM and new awakenings around sensitivity and neurodivergence among women, the tide has shifted such that women are becoming emboldened to take ownership of their minds and lead accordingly. From design and architecture to therapy and publishing, sensitive neurodivergent women are now "coming out" in scores, and we are all better off for it. The narrative is changing, and our job is to keep pushing it further—in media, boardrooms, and classrooms and in our relationships and families.*

## Bias and Shaping Tech

Our neurodiverse future is not limited to those whose expertise is in mental health or the study of the brain. The fields of artificial intelligence (AI), virtual reality, augmented reality, and sensory design are all ripe for neurodivergent input. Eugenia Kuyda developed Replika, an AI chatbot that responds realistically to those in need of listening, empathizing, reassurance, and connection. She developed it as a way to process her grief after her best friend was killed in a car accident. I witnessed a demo of her product and was shocked to see the depth of conversations the bot was able to engage in. This particular bot was actually programmed using the text messaging and correspondence from the friend who died, and so thousands of people were able to say good-bye and privately interact with the bot whose language mirrored that of their dead friend.

Applicable to all of us who go through periods of loneliness or isolation, I also wonder what new insights and applications may emerge if a bot were to be programmed based on the communication patterns of a sensitive neurodivergent. And what if those patterns were studied? Would we be better understood by the larger public? I know this task will be taken up somewhere in the world any day now, and you better believe I hope it's a neurodivergent woman at the forefront.

A *Forbes* article written by entrepreneur Rediet Abebe details a similar concern that AI reflect racially diverse voices. "If teams that set research directions, write algorithms or deploy them are made up of individuals with similar backgrounds and experiences, then we will end up with research that is to the benefit of a similarly narrow and already privileged subset of society," she writes. "The researcher gets to set the question, decide what datasets to use, how to conduct analyses and how to present results."

Entrepreneur Mia Dand of Lighthouse3 uses her media platform to elevate women of color in AI ethics, and Sundance Film Institute New Frontier Lab director Kamal Sinclair is a leading force in helping ensure that technology and virtual reality steer all toward an unbiased future. Sinclair is also founder of a transmedia project called Making a New Reality. The idea moving forward is that if AI will be responsible for decision-making in medicine, hiring, and other aspects of our lives, we want the programming to be clear of bias and include all racial and ethnic groups, genders, and neurodivergents.

An article appearing in *The Atlantic* titled "Beyond the Five Senses" written by Matthew Hutson further expands on additional senses, what happens when they get crossed, and the implications. He writes, "The brain is surprisingly adept at taking advantage of any pertinent information it receives, and can be trained to, for instance, 'hear' images or 'feel' sound." I want to see more sensitive neurodivergent women at the forefront of such work. When I spoke with Stanford neuroscientist David Eagleman about people trading umvelts—so that people can experience the sensory challenges of others—he asked me why I thought someone would want to do that. To which I responded that I don't think many family members, friends, clinicians, therapists, and others would hesitate at the chance to experience and better understand what it's like to be inside an autistic mind and body, for example. When I think of the potential of technology to help all of us better understand and empathize with the experiences of those who are different, I literally get chills.

## Sensory Sensibilities

In the world of architecture, several designers and researchers have written about the application and benefits of autistic sensibilities to all of humanity. Modern architecture in particular, with its minimalist aesthetic, is thought in some cases to have sprung from particular people throughout history who may have been autistic, such as Le Corbusier. "Le Corbusier's designs are a likely response to his atypical brain," Ann Sussman and Katie Chen write in a controversial article. The Nebraska-based firm Forte Building Science published a report titled "Why Buildings for Autistic People Are Better for Everyone," detailing compartmentalization, lighting, moods, comfort factors, and more. Shrub Oak International School in New York, for example, was designed and built entirely according to autistic sensibilities—from the lighting and colors to size and compartmentalization of each room, prioritizing uncluttered open space and avoiding fluorescent lighting. An *Architectural Digest* article about the school points out that "good autism-friendly design adheres to many of the same principles as good design in general: that design elements should have meaning, that an overly colorless and bland space is just as bad as an overstimulating one, and that it's always helpful to consider how spaces make people feel."

And again, all of this innovation points to the urgent necessity of applying neurodivergent thought to diverse fields so that our needs are better understood, accommodated, and—most important—*integrated* into the fabric of life such that we are not an afterthought. We cannot afford to lose any more sensitive neurodivergent individuals to the perils of shame, doubt, unemployment, low sense of self-worth, or suicide.

New innovations in building and design are popping up in a variety of venues, like the NBA's "sensory rooms," which include soothing colors and sounds. A 2018 article about them points out that "these elements help to create a peaceful space away from the noise of the arena. The rooms also feature tools like multisensory play therapy and technology stations. In addition to sensory rooms, the arenas will also have staff members who are trained to recognize sensory needs and provide those accommodations for kids and adults who need them." Not only does this speak to sensitivity accommodations going mainstream, but I'm so glad they mention adults—that seems like progress.

Forward-thinking sensory designers and sensory audiologists involved with the Natural Sounds and Night Skies Division of the US National Park Service's wilderness program in Alaska also hold the power and potential to pay attention to neurodivergent sensibilities. A team of researchers, photographers, and pilots have made their way into remote snowy peaks and wild valleys to capture the sounds of birds and rivers and wind—to simply document soundscapes that are absent of human noise. My own sensibilities make me feel immersed, at one with the pulse of the wild, and the sensations reverberate through me. I would love to see my fellow neurodivergents capturing such sounds, feelings, and experiences out in the world and helping the rest of us benefit from them.

## Expanding Our Senses

By training, Lore Thaler is a psychologist at Durham University in the UK. Her background is in visual perception and how it guides movement, and during her postdoc she became interested in echolocation. This is a form of navigation now being explored by blind individuals—such as well-known figure Daniel Kish, whose YouTube videos first caught Thaler's eye—whereby they use the "clicking" of their mouth and tongue to make their way through streets, homes, and other environments. Thaler has been researching echolocation for more than nine years and uses behavioral experiments and neuroimaging to investigate such things as how people echolocate, what acoustic cues they use, and which brain areas are involved. Researchers like her also study how the brain changes with echolocation, as a person learns the new skills—a process called neuroplasticity.

"What do we consider normal?" Thaler asks me. This is exactly why I wanted to speak with her. Although her research has nothing to do with autistic or highly sensitive women or similar traits, she is essentially researching the act of *neuroqueering— the altering of human performance in the world*. The blind men and women she studies are altering their performance in the world, countering dominant expectations and norms. Clicking is neuroqueering. Or maybe we should call it *sensequeering*. But going further than that, echolocation shows us what is possible when we alter or vary any way that we respond to sensory stimulation or lack of sensory stimulation. So for the sensitive person, what behaviors can we adopt, and likewise what are changes that the world around us can adopt? Some supermarkets have quiet hours for autistic shoppers, for example. What else is possible? What if deep pressure hugs became the norm? What if all offices had Snoezelen rooms to accommodate meltdowns and sensory overload? What if whispering became the communication style at office meetings?

"Sometimes people have very low expectations of people who are blind," Thaler tells me, and I ask her whether she feels it's part of her job to push back on social norms. "Absolutely," she answers. In addition to her research, she and her colleagues conduct workshops to share their research and help other professionals to teach echolocation. "People are taking in this information, and it actually helps change things."

I ask Thaler about other progressive sensory research and changing viewpoints around social norms. "I have colleagues who work with people who are hearing voices," she says. "Traditionally, hearing voices was considered out of the norm, an 'abnormality,' and they've demonstrated that almost everyone at some point in their lives has this experience and that there's a spectrum." Many researchers at Durham University—and in the UK more broadly—seem to be taking expansive stances in their research, conclusions, and recommendations for practice. A large percentage of the research I came across while researching this book was from the UK and other European countries, such as the Netherlands and Italy. "The idea in the UK is to integrate," Thaler says. "There used to be special schools for the blind or those with learning disabilities, and now the idea is to integrate, rather than separate. Sometimes that can go wrong, but the idea is to try to integrate."

Sometimes people ask Thaler why she's interested in this area of research, and she says it's a powerful paradigm shift to learn about how perception changes. "Not many people are working in this area," she says. And looking to the future, as more scientists, designers, and inventors cater to sensory differences, she says there is the problematic question of imposing, rather than collaborating. "Whenever you introduce assistive technology, one has to be careful that people actually want to have it. If people find it useful and they want it, I'm all for it." She points to hearing aids as an example; many people asked for that kind of functionality rather than outsiders imposing the technology.

Regarding innovation in research—especially how animals play a role—Thaler says, "Much of the literature on echolocation comes from bats, and I use it as inspiration." Although she doesn't work with bats or other animals, she gets to explore what is similar between species. If bats took thousands of years to evolve their ability for echolocation, I ask her, what is the future for humans? What kind of sensory perception or communication is in our future? "I think it's a fundamental question," she says. "As a human you have your sensory surfaces, like skin and pressure and temperature, hearing and sight. And with a new sense, you'd still be limited, but obviously you can get different information. If you use echolocation instead of hearing, you get information that is different. So is this now a new sense?"

What I find so intriguing about Thaler's research is her eye on the long view of humanity's evolution. When I think about it, isn't it possible that sensitive neurodivergent folks in another era could actually be considered advanced for being so affected by sensory stimuli? And what new ways of operating and interacting might sensitive folks come up with were they given full range to explore and use their sensibilities? Could their sensibilities create cities and towns that are inherently soothing? And would living in such ways help solve other societal ills, such as aggression, hostile communication, war, greed? Or might they come up with forms of communication that don't rely so much on sound, just as blind people are using echolocation instead of sight to navigate space? Replacing sight with echolocation allows people to absorb similar informational input but use a different sense, and this helps them function in the world with less distress. What is the equivalent for those of us overwhelmed by noise, for example?

Bill Davies, the autistic acoustics professor from the UK we met in Chapter 5, wants to get designers, planners, architects, and engineers in a room to better understand how poor acoustics affect and marginalize people and says that increased interest in concepts like soundscape ("which proposes a more nuanced and complex understanding of human response to sound environments") would be helpful. "The cocreated design process seen in soundscape projects, where users/residents/stakeholders are involved in the design from the earliest stages, would certainly be useful in making more accessible environments," he says, and adds that getting the initial design right is much less expensive than trying to redo something after it's been built.

Davies also wants to see more basic research on autistic auditory perception, especially research looking for strengths and differences, as opposed to "deficits." And like many people, he wants to see more autistic individuals at the helm—planning the research questions, conducting the research, interviewing other autistic people. "I think we might learn a lot if researchers started simply by listening in detail to autistic people's accounts of their everyday auditory experiences. I find that, with a bit of prompting in the right environment, most people (autistic or not) can give a rich account of their experiences with sound." Davies is encouraged by the increase in participatory research, for example, and wants to see more of it. "This mirrors the cautious progress toward patient involvement in medical research more broadly," he says. "There are small pockets of genuine cocreation, where established neurotypical autism researchers have entered into open research collaboration with autistic individuals. Younger autism researchers seem to be more open to this, and the growing band of autistic PhD students are producing exciting work."

Joel Salinas, the Harvard neurologist who has synesthesia, speaks widely about embracing human difference—digging deeply into how much love we could all be sharing by acknowledging the mind-boggling scientific understandings of how different and unique each individual is. He also mentions to me what it would be like to manifest that love, understanding, and acceptance in a way that leads to concrete change in how the world operates and is structured. "What's vital is being open to our differences, just as we are open to our similarities," he says.

Salinas also says that "we need to create environments that are more customizable and less likely to cause stress—environments where people have options." For example, airports have designated smoking areas; so one might imagine that airports could also set aside places for people who are highly sensitive to environmental stimuli—"environments to find your own personal Zen," as he puts it. There are also other considerations, such as noise-canceling headphones, visors, particular clothing, and more. Salinas reminds me of the example of left-handed scissors, which have met a wide need and don't impose a burden on all the right-handed people in the world. Salinas is keen on collaboration and reminding ourselves that we all share this world and we can work out what everybody needs while trying to minimize the burden to others. "If we want to survive together, we have to collaborate a little bit," he says, referring to managing differences with compassion and empathy.

Occupational therapist Teresa May-Benson—whom we met

in Chapter 4—echoes what many in the field of sensitivity have known for decades but what is just on the cusp of being implemented. "We need to educate medical professionals, psychologists, the justice system, and other professionals about the importance of sensory processing and how it impacts behavior," she says. An episode of the Netflix show *Atypical* called "In the Dragon's Lair," for example, details a scene in which the main autistic character's sensory overload and corresponding panic look to a police officer like "disobedience." In the show, the police officer later gets some basic introductory information about autism to prevent future misunderstandings. So much information and education is available that is not getting out to those who need it—those in power and their constituents—and so the acts of reframing such narratives and *telling different stories* around sensitivity and behavior are not just revolutionary, but also very practical.

It would be wise for parents, teachers, spouses, colleagues, police officers, therapists, doctors, and others to learn this information and make adjustments so that sensitive neurodivergents can thrive. It shouldn't be about coercing or "retraining" neurodivergent folks to be different from who they are, but rather about all of us—neurodivergent and neurotypical—making room for everyone. As we know by now, the lifelong effects of masking can be dangerously draining and hazardous to emotional, physical, and spiritual health.

## Spanning Continents and Diagnoses

The coming changes are not limited to the US or UK, or to one cluster of diagnoses. Mariana Garcia is a therapist in Mexico who originally wrote to me after reading a piece I produced with Elaine Aron. When I announced my first Highly Sensitive & Neurodivergent retreat, she was the first to register, and we stayed in touch and corresponded about her hopes for such work in Mexico.

"I was interested in this field when noticing two things," Garcia says. "The number of adults who suffer from social anxiety caused by having felt maladjusted throughout their history—I realized that they shared several characteristics such as a taste for reflection and emotional intensity. I found that the neurodiversity and high sensitivity lenses both made sense with what I heard from friends, family, and patients." Garcia also observed a disconnect in schools: "Many of these children feel overwhelmed and afraid, which becomes a vicious circle that affects their emotional, cognitive, and social development."

Garcia plans to introduce the neurodiversity framework in Mexico, with a focus on high sensitivity, so that "instead of children growing up feeling inadequate and ashamed for not being 'normal,' they can know themselves and we as adults can create the necessary conditions to help them develop their potential." She wishes society at large could be aware that there are different minds and ways to create contexts in which highly sensitive and neurodivergent individuals can "self-realize." Ending shame and stigma is a huge focus in Garcia's current practice. She notes that language in particular is needed in Mexico to help better reflect the internal worlds of sensitive neurodivergent people. "Deficit" is still the common viewpoint.

Like others interviewed throughout this book, Garcia sees the challenge of integrating the neurodiversity framework into the current medical system in Mexico. She says that conventional doctors and therapists should not be quick to judge neurodiversity thinking as dismissive of treatments or attempts to help; someone who is a neurodiversity advocate can *also* be open to benefitting from certain therapies to improve their lives.

Like many other advocates, Garcia's focus is on integration. The idea moving forward within psychology and related fields is to continually revise and update notions of human nature and behavior and the role of societal norms. "The world will benefit significantly from talents such as empathy, emotional intensity, certitude, sensitivity, ability to detect details, depth of thought, will to embrace, and many other things that we need in a time where alienation, coldness, superficiality, and emotional hardness are predominating," Garcia says. She, like many others, does not want her profession or those influenced by her profession to take an extreme or fanatical viewpoint. She wants to open doors and push forward a new way of thinking, while at the same time not denying the various experiences of sensitive neurodivergent people. As we've already seen, neurodivergent individuals can still embrace medications and therapies while also taking pride in their identity.

It's clear we need a massive infusion into all fields of sensitive neurodivergent folks, especially women, who are too often barred from success because of gender bias, mental health stigma, and stereotypes about mental differences. We also need to look more closely at the subject of sensitivity in other categories, such as what is currently called borderline personality disorder, schizophrenia, OCD, and bipolar disorder. Andrew Solomon's 2012 book *Far from the Tree: Parents, Children, and the Search for Identity*, for example, illuminates how parents navigate various types of differences in their children. One day I came across a video from Yale University about Solomon and his friend Elyn Saks, both Yale alumni and both of whom had begun sharing openly about their mental health challenges—Solomon's depression and Saks's schizophrenia. An Oxford graduate, Saks started experiencing schizophrenic episodes in her twenties and developed a robust career around mental health law, human rights, and policy. She is also a MacArthur Fellow and started the Saks Institute for Mental Health Law, Policy, and Ethics at USC, where she is a professor.

She is now working on supportive decision-making efforts to help involve individual-appointed family members or friends to help intervene during mental health episodes and has been in discussions with Johns Hopkins University, Columbia University, and other stakeholders in the medical system. Saks wants people to be the "architects of their own lives." She says she is pro-psychiatry and anti-force. When I talked with her, I got a vision of the future for neurodivergent women: open, accepting, present, capable, accomplished, determined, focused.

Saks tells me how hard it has been for graduate students, for example, to disclose their conditions. "In recent years, when we would first meet as a group, only one person would raise their hand [to identify herself or himself as having a mental health challenge], which is stunning. But this year, 75 percent of the group raised their hands to disclose, including a woman who had never said it out loud to anyone aside from her family." She continues, "I want to see more research into what works, stigma, why people aren't getting care, and ways to get people to want treatment so they don't use force."

Saks understands the neurodiversity framework and other movements such as "Mad Pride," and she respects such approaches. For her personally, medication and therapy work, and she therefore supports their use for others if they choose, "but it's an individual choice for everyone," she says. She views her schizophrenia as a "biochemical illness that needs medication and therapy in order for me to do well." It's a pragmatic approach, she says, and if others can do well seeing it differently, that's fine by her.

Saks is clearly aware of the challenges within psychiatry, and it's refreshing to hear someone acknowledge various perspectives. She tells me, for example, that she thinks "psychiatrists are doing a great disservice when they tell us to immediately lower our expectations, because with proper resources and support, people can live up to their potential." Plenty of people living with schizophrenia are doing well, she says, including friends and colleagues of hers who are doctors, lawyers, and academicians. Workplace acceptance has been critical for her, and USC accommodates her well. For example, since she finds teaching stressful, someone teaches her classes while she supervises students one-on-one. She says, "Work is incredibly meaningful in terms of feeling productive and valued and doing something good."

Saks is involved in a study that is surveying clinicians, patients, family members, and friends "about a new word for schizophrenia because it's just so misunderstood." This points to the importance of language and reframing. How do we talk about difference in our families, schools, and workplaces and in government? Saks and I chuckle when we agree that this effort is essentially about *branding*.

Regarding family and friends, I ask Saks about her personal ecosystem. Who's necessary in her immediate inner circle? I know from my own experience that a "support system" for us neurodivergents is often different from a typical support system of a casual friend here and there. "I wrote a list of people who have helped me—my psychiatrist, psychoanalyst, cardiologist, cancer doctors, lawyers, housekeeper, friends, and husband." She tells me, "Society is getting more understanding, and there's more openness, even in the media. People are telling more positive stories."

## Academic Research: Belly of the Sensory Beast

Change is also coming from within academia. Hailing from Kentucky, B. Blair Braden got her PhD from Arizona State University, Tempe, and now leads the Autism and Brain Aging Laboratory on campus. With a background in behavioral neuroscience, she realized the need for closer study of autistic adults because so few researchers were focusing on individuals older than eighteen, let alone women. When she and her colleagues made their initial grant application to the Department of Defense, they needed their participants to be as similar as possible and focused on men since more men are diagnosed. That didn't satisfy her, so she got another grant from the Arizona Biomedical Research Commission to do the same study in women; however, she hasn't gotten enough research subjects to be able to publish the results yet. "But another grant from the National Alliance for Mental Illness will allow us to continue to follow both sexes for the next four years," she says.

Braden had been recruiting women ages forty to sixty-five for a year when she and I spoke. Across both sexes, the vast majority of her participants were not diagnosed until adulthood. "There's such a large difference between the sexes in terms of diagnosis rates," she says. For most participants, autism did not even get added to the DSM until after they had been born and often after they were already out of elementary school.

Braden points out the challenge that masking plays from a research perspective. Let's say a woman goes to a therapist or researcher and answers questions the way she's used to doing at work or at a coffee shop—using all the neurotypical nuances, cues, behaviors, gestures, and other ways of interacting that she has learned to use. An observer will find it difficult to recognize that the woman is neurodivergent. This is especially an issue for women who don't even *know* they are masking or who have been doing it for so long that they think it's normal to feel exhausted and anxious after most interactions and conversations. "There's finally some more research coming out about how women are better at camouflaging, so our tests may not even be sensitive enough for girls or women," Braden says, meaning that standard diagnostic criteria don't take into account masking, so women slip under the radar. "We're still grappling with human cognition and traits. The way things are laid out in the DSM right now may not be the most accurate."

The goal of Braden's research is to identify challenges for autistic individuals as they age and what can help them in that aging process. She and her colleagues are now developing interventions to address anxiety, depression, and executive functioning abilities. For the future, she wants her research to inform what the standard of care should be. "We're particularly concerned about so-called high-functioning adults who may have achieved a certain level of independence, who can work and support themselves, living independently—but if they are affected by aging a little early, then those changes are where we want to come in and consider what we can provide to help keep this person independent."

As we've seen, much more work is being done in Europe, but even there, little work is focusing on women. Reconsidering diagnostic tools is a priority for the international autism research community, says Braden, especially since boys and men are diagnosed far more often. There's a sense that girls and women are getting missed, so changes in testing are needed.

"I think we give a lot of leeway to kids—that it's okay for kids to be different," Braden says. "But once you're an adult, it's not okay. I'm really passionate about advocating that it's okay for people to be different, and it should be celebrated." In her research, she says, "We realized how much peace it brought these adults to finally find out they're autistic. Acceptance is a huge piece."

In sum, Braden says, "Doing this research has totally changed my life. Working with adults on the spectrum has given me such an appreciation for how different we all are. There's really nothing right or wrong about people; we're all just people doing our best. I have become an infinitely more accepting person by getting to know people who are very different from me."

## My Parting Thoughts

While I was conducting interviews for this book, many people asked along the way about why we need to categorize or diagnose at all if there are so many similarities among synesthesia, autism, HSP, SPD, and ADHD. As we know by now, categorization and diagnosis are largely functions of the DSM, insurance companies, and doctors and therapists needing to use such labels to find support, treatment, or therapy for individuals. It's possible that one day all of them will be in the DSM, or it's equally possible that one day none of them will be in the DSM. And as Joel Salinas reminds me—even for something like synesthesia, outside of research centers—what are available for the public, at best, are self-reporting questionnaires. On the other hand we've seen how for some people, getting the "correct" diagnosis and confirmation from a doctor felt important and affirmative. Perhaps it depends on your work and how close to the material you are and what you think is necessary and how you approach experts and "expertise." Indeed, I've heard other stories of women trying to get diagnosed but finding physicians to be clueless, misinformed, and not up to date on the latest studies such that the "expert" in the room is the patient herself.

After all, what does it even mean to ask these questions or to wonder whether you "have" ADHD or you "have" autism? What labels and diagnoses mean today is different from what they will mean five years from now, which will be different twenty years from now—because they are not static; rather, they are floating concepts created by humans that morph and change over time. So who gets to decide? Many people, especially women, are done living according to the categories defined by others. But other people find those categories to be necessary for them to access life-saving treatments, support, and accommodations. How do we straddle such a wide array of needs that fall under one name, label, or categorization?

It's a confusing, necessary question, and for me the answer—in addition to acceptance—has been to affirm the similarities we all share as sensitive neurodivergent women. Through those similarities, our community—our tribe—is that much bigger. I know allies who are nonspeaking or wheelchair-bound or heavily medicated or in and out of treatment homes. I can stand next to them, march with them, hope that the privilege in my life can be used to support them—and in the same breath I can affirm that we both dislike bright lights and loud sudden noises and that migraines suck.

Once I embraced my own labels, categories, and identifications—I mean truly embraced them and got my family, friends, and colleagues on board—I almost didn't need them anymore. Once I learned about them, I adjusted, we all adjusted, and now I have a thriving life, meltdowns and all.

The picture that begins to emerge is that humans come in so many different flavors that the categories we've defined potentially fall away. This is not denying the importance and utility of such categories, but it questions their primacy, their fixed qualities, and the ways in which we employ them in intimate conversations and wider cultural contexts. It's important that we see differences, that we don't deny them; but let it stop there and respond to everyone with kindness and help—pure, unadulterated kindness and help that are not premised on whether someone is "high functioning" or "low functioning." We know enough about masking and camouflaging to *not* ignore the reality that someone with a job, salary, and family may be steps away from suicide, just as clearly as we know that someone who does not speak may need assistance at school to communicate. Underneath, we are all more similar than we realize, but we don't talk about it, so no one knows it.

What ultimately worked for my own growth, acceptance, and healing is fourfold: (1) finding the right career; (2) coming to understand my needs; (3) communicating those needs and having them respected by friends, family, and colleagues; and (4) learning more about my body. I started noticing near the end of the research for this book that I experience the same calm, serene feeling reading new reports and studies about how the brain and body work as I do when I read Sufi poetry or the great philosophers, artists, and intellectuals of our time. Both fill me with a reassuring and settled feeling as I take in the contours of our human design and experience.

The evening when I found the article detailing "interoception-focused therapy," the name immediately resonated with me. When I saw that the research was being piloted by a female neuroscientist who was trying to understand and decrease anxiety among autistic study participants, I knew I needed to pay attention. She had found a buried paragraph in a report indicating that autistic people had a difficult time detecting their own heartbeats and that may contribute to their anxiety. What she read also indicated that autistic people had more visceral responses to other people in pain than a control group, indicating heightened empathy. Reading just these few sentences about Sarah Garfinkel's research journey and the clues she was noticing made me feel drawn to her because I knew she was onto something.

The next morning, I did jumping jacks—and was able to feel and count my heartbeats without holding a finger to my pulse. What followed were vivid images about the insides of my body—my cells, organs, muscles, brain stem. I had been craving a more precise and refined mental map of my physical insides. I felt that same deep feeling of being seen. Being recognized. And this is what Garfinkel's research was indicating was necessary to help reduce autistic anxiety as well.

So get to know your body. Search online for documentaries about how the body works. The information may ground you and help you feel like you haven't felt before. Or read more about and try some of the other approaches discussed in this book—occupational therapy, Integrated Listening Systems, Snoezelen design. We're not taught to be attuned to our sensory selves.

Another central healing balm for me has been my work. For a long time I felt cooped up in my brain with ideas and thoughts and sensitive reflections that didn't have an outlet. Shame started to dominate, especially as I tried to find work that met my inner world. Once I was able to take a stand and share what my inner world was really like, then I felt I could finally exist and place my feet on the ground of the outer world. My brain "turned on" again, and my old knack for mental mapping of people and their interests, stories, and internal lives came back online and I could really get to work. The Neurodiversity Project was born, linking together these interests, speakers, ideas, deeper conversations, and real-world effects with systemic change inside of people and the institutions and organizations where they work, including hospitals, universities, schools, media and tech companies, and therapists' offices.

It's hard to make your needs known—especially when you may not yet understand the full spectrum of those needs. It's challenging work that must be tackled with incremental steps. If you aren't yet accustomed to asking a friend or partner to pause a conversation because the content is overwhelming you, or explaining to family members why you can't go grocery shopping at certain hours of the day, start with what feels most doable to you. Perhaps your friend already has some idea of your sensitivities, so she may not be surprised at the request to change up the conversation. Perhaps your partner sees that your grocery shopping is more effective when you go in the morning rather than at night. And don't worry if anger comes up when first articulating, voicing, and setting boundaries around such needs. You are flexing a new muscle that takes time to strengthen, and the people in your immediate circles can shoulder their own feelings in order to support you with love during a time of growth.

By the end of my research for this book, I felt so *seen*. And relieved. Being able to learn, digest, process, and ultimately implement and integrate so much information has been a healing process for me. Perhaps the most important thing I learned from researching and writing this book is the value of acceptance. With acceptance comes accommodation, understanding, and a sense of spaciousness from others and society at large that allows neurodivergent individuals to develop and grow in their own way and ultimately identify how to plug their strengths in to neurotypical settings that once felt uncomfortable. Acceptance is at the core of what then enables people who feel marginalized to take risks, expand their sense of belonging, apply themselves in work and relationships, and thrive.

## The Path Forward

The way we do medicine and the way we talk about sensitivity and difference as a society need to be reframed. Medical schools, police associations, writers of the DSM, professors and scientific researchers, schools and parents, human resources departments, innovation offices housed within companies—all of them need to be included in a larger conversation. This is not about autism or ADHD or women or men—this is about the fundamental way in which we view, handle, and talk about difference and how we empower or disempower people. There are now hospitals in Canada, for example, that refer patients to art museums as "treatment" for depression and anxiety; there are virtual reality programs that help others experience the sensory world of an autistic child; there are therapists acting as first responders for the homeless rather than the usual police squad.

Within research and the medical system, there are a host of women speaking out and writing on gender bias and how to do medicine better—such as we've seen with Maya Dusenbery's book *Doing Harm* and Angela Saini's *Inferior*. Emergency medicine doctor Shannon McNamara writes and speaks about the intersection of gender bias, queerness, and emotional labor for the media platform FemInEm. Dr. Rana Awdish speaks out on the forgotten importance of emotional connection and resonance in her 2017 medical memoir *In Shock: My Journey from Death to Recovery and the Redemptive Power of Hope*. Lissa Rankin, whom we met in Chapter 2, continues to write and speak widely on building a medical system that prevents burnout and honors science, ancient wisdom, and areas of practice that science and medicine should be looking at but haven't yet devoted the time and money to. And the psychiatrist Alexandra Sacks is reframing and depathologizing the conversation about and experience of new motherhood and the emotional roller coaster that ensues. We need more conversations on all of the above.

The field of medicine needs an overhaul. With the number of people experiencing loneliness on the rise, more people become ill and turn to doctors, who are thus put in the position of needing to figure out people's social lives. But doctors are getting burned out and also need support and others to turn to. And the cycle goes on. Without people being able to open up, share, and connect with others about their internal lives, nothing will change. People fear being exposed, thinking to themselves, "I don't want people to know my challenging parts," so they hide and stay separate and isolated. And the isolation turns into physical and psychological symptoms.

It's not enough to say we need more connection or to point out that loneliness is a problem. We need to know *how* to connect, which means learning how to have better conversations with others and reveal ourselves to others more fully. Sharing our struggles, especially our mental health ones, is a direct path to connection. Not everyone will know how to respond or what to say, but by opening your door, you help open theirs. It can take time, but soon everyone in your circles of family, friends, and colleagues will feel a little more at ease being themselves. The amount of stress decreases, and we are healthier—and more connected—as a result.

There is so much work to do moving forward. Through this book you've been given a glimpse into what some remarkable women and men are doing in their unique niches of work. At their core is an unapologetic commitment to authenticity and "unmasking"; it takes bravery and determination to weather the storms that can ensue. But after the transition, you get to *lead your own life*. I know from personal experience that so many people wish for this, but they feel trapped and don't know how to make it happen, or they don't have the language to imagine a better way. I hope this book has given you that language. It's time to "come out," and as you do, what once felt like a hidden alternate sensory enclave blooms to become an enveloping universe that feels like home, that you've longed for, and that you embrace with awe.

# Acknowledgments

First, I thank my editor at HarperOne, Hilary Swanson. Thank you for saying yes to tea and to this book. Your support and enthusiasm have warmed my heart, and I'm so grateful to have worked with you.

Overwhelming gratitude to the many authors I've worked with through The Neurodiversity Project—Pico Iyer, Bill Hayes, Gabor Maté, Lissa Rankin, Joel Salinas, Maya Dusenbery, Steve Silberman, Rev. angel Kyodo williams—thank you for saying yes to something so new and different. I hope our time together has enriched your lives just as you have enriched mine—and thank you for cheering on this book.

I also thank the many authors I've been fortunate to interview through my work with the UC Berkeley Greater Good

## Acknowledgments

Science Center—Krista Tippett, Roshi Joan Halifax, Courtney Martin, Sebastian Junger, and many more.

Thank you to my mentors Susan Cain, Elaine Aron, Nick Walker, and Rev. angel. You four have been so pivotal in my work and development.

Thank you to the many creators who participated in our first #Reframe Conference just as I was finishing this book! Kamal Sinclair, Shawn Taylor, Nidhi Berry, Liz Fosslien, Casper ter Kuile, Tina Sacks, and so many others—it has been a joy to collaborate and cross-pollinate intersectional ideas.

Thank you, also, to the many volunteers who have helped to make each event happen with such excitement and meaning, connection, and belonging.

To the whole HarperOne team—you've been so kind and supportive from day one. Thank you for your tenderness, and unlike many publishing teams these days, we've actually had face-to-face time, which I'm so grateful for!

To the many communities who held me as I journeyed through—Thrive East Bay, On Being, The Aspen Institute, The Center SF, The Family Spirit Center, OZY Media, 101 Surf Sports, Heart Source, and the Tahirih Justice Center.

Thank you to my early high school and college teachers who encouraged my voice.

Thank you to the many amazing women interviewed for this book—thank you for sharing your stories and opening up your hearts with me. I know how hard you work every day—I see you, I feel you, I thank you, and I am sending fire your way to light you up as you release your gifts to the world.

To my friends who've known me since childhood and gracefully danced this dance with me: Kate, Shirin, Lydia, Becky.

To my extended family of in-laws, aunts, uncles, cousins, colleagues, and family friends across the US and abroad, thank you for your love and gentleness.

To my parents, thank you for exemplifying how to thrive and fight for sensitivity, each in your own way.

To my siblings, my protectors, I am so grateful for our tribe.

To my partner, thank you for standing alongside me on this journey and for your patience and support. And thank you for always keeping things fun. :)

To my daughter, I hope for you a world where the expression, acceptance, and flourishing of sensitivity becomes the norm. Your perceptiveness and humor are gifts beyond measure and I love you. Thank you for your patience as I have slowly put the pieces of my own story back together.

# Resources

## MEDIA RESOURCES

I find media, and particularly film, to be a powerful medium for re-framing societal narratives surrounding divergent identity and mental health. I hope the below recommendations stir your imagination for what is possible in your own life. Breaking stigma and shaking up the conversation in our communities is how change happens and I can't wait to see what you do. Tell me about it! If you'd like to share, you can comment on Facebook at @neurodiversityproject, tag me on Twitter at @bopsource, or send me a note through divergentlit.com.

### CAPTAIN FANTASTIC

It's unusual for Hollywood to take on the mundane aspects of mental health, and this film does some of that, depicting how an unconventional family struggles to balance its philosophical convictions about modern life with the reality of modern life. It's a beautiful film, portraying loss and grief juxtaposed with suburban indulgence and the pull of nature.

## THE HOURS

One of my favorite films from when I was younger, this movie woke something up within me, but I wasn't sure what at the time. Now, of course, I understand that its precise portrayal of women and masking from a historical and contemporary perspective is something that many can relate to. The way the stories are interwoven also creates a kind of synesthetic feeling.

## ATYPICAL

The Netflix show does a wonderful job of capturing the experience of sensory overload and also illustrating how neurotypical friends and family coexist with and work alongside to create a sense of community with neurodivergents.

## THE LAST BLACK MAN IN SAN FRANCISCO

Portraying my hometown of San Francisco, fellow School of the Arts alum Joe Talbot captures a felt sense of the city's culture and streets. I particularly appreciate the way the young playwright, Mont, is depicted—clearly neurodivergent, his sensitivity, empathy, artistic sensibilities, and kindness are portrayed with such depth by actor Jonathan Majors.

## FROZEN 2

This breathtaking animated film shows the pull that many have to abandon society's social norms and find community while doing so, something that was a strong motivator in creating The Neurodiversity Project. The film depicts bright magenta and midnight blue as the main character navigates an icy world and literally follows a voice only she can hear. It creates a synesthetic sensation between land and person. It also, in a sense, normalizes the experience of "hearing voices" and depicts the experience in a very grounded, spiritual way, which is a bold and beautiful move for Hollywood. I think this film will greatly impact generations to come.

### LOU ANDREAS-SALOMÉ

This is one of the most gripping portrayals of a woman's intense curiosity, determination, and fierce intellect and the ways in which the frames of society meld together to shape how we view personhood, identity, and ability. It is a fantastic feminist account of Western thinking in psychology and philosophy.

### RUNNING WITH SCISSORS

It's been a while since I've watched it, but this film captures some harder aspects of growing up in neurodivergent families. When mental difference isn't accompanied by kindness and humility, it can turn mean. Many people get into a kind of "guru" mentality, which can turn into narcissism, and the effects on children and families are painful.

### THE GLASS CASTLE

This film also depicts growing up in a neurodivergent family, in combination with the effects of alcoholism and trauma and their lasting aftereffects. The main character finds her own life through writing and establishes herself in a more mainstream neurotypical world but eventually finds a harmony and synthesis of both worlds, by forgiving her father. A sweet film.

### INFINITELY POLAR BEAR

Interweaving themes of interracial relationships, class, biracial identity, and mental health, the film is a tender and realistic portrayal of a family navigating a bipolar father's episodes. Like *Captain Fantastic*, the film takes on the mundane of finances, breakfast, school routines, and more topics that affect neurodivergent families the world over.

## MY ARTICLES AND INTERVIEWS

For further reading, here is a sampling of my work related to mental health, neurodiversity, media narratives, and culture shifts.

### ARTICLES

"What Neurodiversity Is and Why Companies Should Embrace It." *Fast Company.* https://www.fastcompany.com/40421510/what-is-neuro diversity-and-why-companies-should-embrace-it
> This article delves into neurodiversity at work, featuring interviews and recommendations, with a focus on women and Silicon Valley.

"About High Sensitivity, Autism, and Neurodiversity." The Highly Sensitive Person website. https://hsperson.com/about-high-sensitivity -autism-and-neurodiversity
> Elaine Aron called me to ask some questions about how the neuro-diversity framework connects autism with the highly sensitive trait, and this is the blog post we coproduced.

"The Tiger's Journey of Nonconformity and Neurodivergence." *Quiet Revolution.* https://www.quietrev.com/tigers-journey-nonconformity -neurodivergence
> This piece began as a very personal Facebook post to open up to my family and friends and to share what I was going through, and then it turned into a published essay—and then it turned into *Divergent Mind.* A special piece, it gives further historical context on my own life story.

"Why Neurodiversity Matters in Healthcare." The Aspen Institute. https://www.aspeninstitute.org/blog-posts/neurodiversity-matters -health-care
> This is an in-depth Q&A with me about my writing, work, personal experiences, and what led to The Neurodiversity Project.

"The Science of Awe and Why It Matters at Work." *Quiet Revolution.* https://www.quietrev.com/the-science-of-awe-and-why-it-matters-at -work

Here I share some of the research on awe by my colleagues at UC Berkeley, as well as practical examples of how awe can be woven into design and daily life. A sense of awe, wonder, and immersion certainly have sensory components, and the way buildings and workplaces are designed can greatly impact our mental well-being.

"How Design Is Helping Us Understand the Brain." *Fast Company.* https://www.fastcompany.com/3061887/how-design-is-helping-us -understand-the-brain

This piece highlights a TED residency project whereby interactive design exhibits help attendees better understand neuroscience.

### INTERVIEWS

"Finding Stillness in New York City: An Interview with Bill Hayes." Garrison Institute. https://www.garrisoninstitute.org/blog/finding -stillness-new-york-city

Hayes is an award-winning author and photographer and was the partner of the celebrated neurologist Oliver Sacks. In this interview we discuss topics of sensitivity and reflection, Sack's life, finding community and a sense of place, and forging connections across neurodivergences.

"The Fire of Now." Garrison Institute. https://www.garrisoninstitute .org/blog/the-fire-of-now

In this interview with author, activist, and Netflix director of inclusion, Darnell Moore, we talk about Moore's coming-out story as a young queer black man and how he reconciled his various identities alongside his spiritual community. Themes of liberation, contemplation, and societal transformation are all discussed.

"Women Rowing North: An Interview with Mary Pipher." Garrison Institute. https://www.garrisoninstitute.org/blog/women-rowing-north -an-interview-with-mary-pipher

Pipher, a therapist, is an acclaimed bestselling author who traces the impact of wider cultural shifts on her individual clients in therapy. In this interview she takes on the subject of women in their later years, aging, and reframing narratives about what it means to get older.

"Mindful Design and Remembering that Women Are Half of Humanity." Garrison Institute. https://www.garrisoninstitute.org/blog/mindful-design-remembering-that-women-are-half-of-humanity
This interview with author Caroline Criado Perez exposes the ways in which women have been left out of virtually all aspects of life—from urban planning to health care—and Perez issues a challenge to readers to "remember that women are half of humanity."

"How to Address Gender Inequality in Health Care." UC Berkeley Greater Good Science Center. https://greatergood.berkeley.edu/article/item/how_to_address_gender_inequality_in_health_care
This interview with *Doing Harm* author Maya Dusenbery exposes the extent of gender bias in scientific research and the medical system.

"Does Neurodiversity Have a Future?" UC Berkeley Greater Good Science Center. https://greatergood.berkeley.edu/article/item/does_neurodiversity_have_a_future
*NeuroTribes* author Steve Silberman is interviewed here on the impact of politics on neurodiversity policy and research.

# Further Reading

### Introduction

Aron, Elaine. "About High Sensitivity, Autism, and Neurodiversity."
The Highly Sensitive Person, April 26, 2017. https://hsperson
.com/about-high-sensitivity-autism-and-neurodiversity/.

Muzikar, Debra. "Neurodiversity: A Person, a Perspective, a
Movement?" The Art of Autism, September 11, 2018. https://the
-art-of-autism.com/neurodiverse-a-person-a-perspective-a
-movement/.

Nerenberg, Jenara. "The Tiger's Journey of Nonconformity and
Neurodivergence." Quiet Revolution. https://www.quietrev.com
/tigers-journey-nonconformity-neurodivergence/.

———. "What Neurodiversity Is and Why Companies Should Embrace It." Fast Company, May 19, 2017. https://www.fastcompany
.com/40421510/what-is-neurodiversity-and-why-companies
-should-embrace-it.

———. "Women Rowing North: An Interview with Mary Pipher."

Garrison Institute, March 12, 2019. https://www.garrisoninstitute .org/blog/women-rowing-north-an-interview-with-mary -pipher/.

Rabin, Roni Caryn. "Health Researchers Will Get $10.1 Million to Counter Gender Bias in Studies." *New York Times*, September 23, 2014. https://www.nytimes.com/2014/09/23/health/23 gender.html.

## Chapter 2: Reframing Sensitivity

Acevedo, B. P., E. N. Aron, A. Aron, M. D. Sangster, N. Collins, and L. L. Brown. "The Highly Sensitive Brain: An fMRI Study of Sensory Processing Sensitivity and Response to Others' Emotions." *Brain and Behavior* 4, no. 4 (2014): 580–594. doi: 10.1002 /brb3.242.

American Occupational Therapy Association. "Frequently Asked Questions About Ayres Sensory Integration," 2008. https://www .aota.org/-/media/Corporate/Files/Practice/Children/Resources /FAQs/SI%20Fact%20Sheet%202.pdf.

Aron, Elaine. "Are You Highly Sensitive?" 1996. http://hsperson .com/test/highly-sensitive-test/.

Cassani, Monica. "Language of Mental Illness 'Others' People: It's a Human Rights Violation." Mad in America, September 19, 2017. https://www.madinamerica.com/2017/09/language-mental -illness-others-people-human-rights-violation/.

Iacoboni, Marco. *Mirroring People: The New Science of How We Connect with Others.* New York: Farrar, Straus, and Giroux, 2008.

Rankin, Lissa. *The Anatomy of a Calling: A Doctor's Journey from the Head to the Heart and a Prescription for Finding Your Life's Purpose.* New York: Rodale, 2015.

*Sensitive: The Untold Story.* Directed by Will Harper. The Global-Touch Group, 2015.

## Chapter 3: Autism, Synesthesia, and ADHD

Acevedo, B. P., E. N. Aron, A. Aron, M. D. Sangster, N. Collins, and L. L. Brown. "The Highly Sensitive Brain: An fMRI Study of

Sensory Processing Sensitivity and Response to Others' Emotions."
*Brain Behavior* 4, no. 4 (2014): 580–594. doi: 10.1002/brb3.242.

Aron, Elaine. "About High Sensitivity, Autism, and Neurodiversity."
The Highly Sensitive Person, April 26, 2017. https://hsperson
.com/about-high-sensitivity-autism-and-neurodiversity/.

### Chapter 4: Sensory Processing "Disorder"

Allergic to Sound. "Understanding Misophonia, Misokinesia, and
Sensory Processing Disorder." https://www.allergictosound.com.

iLS: Integrated Listening Systems. "The Safe and Sound Protocol:
What Is the SSP?" https://integratedlistening.com/ssp-safe-
sound-protocol/.

Kanter-Brout, Jennifer Jo. "What Is Misophonia?" Star Institute for
Sensory Processing Disorder. https://www.spdstar.org/basic
/misophonia.

STAR Institute for Sensory Processing Disorder. "About SPD."
https://www.spdstar.org/basic/about-spd.

### Chapter 5: Well-Being

Devlin, Hannah. "Autistic People Listen to Their Hearts to Test Anti-
Anxiety Therapy." *The Guardian*, December 14, 2018. https://
www.theguardian.com/society/2018/dec/14/pioneering-therapy
-for-autistic-people-with-anxiety-undergoes-clinical-trial.

Porges, Stephen W. *The Polyvagal Theory: Neurophysiological Foundations of Emotions, Attachment, Communication, and Self-regulation*.
New York: Norton, 2011.

### Chapter 6: Home

Ehrenreich, Barbara. *Dancing in the Streets: A History of Collective Joy*.
New York: Metropolitan Books, 2007.

### Conclusion

Adams, Seth. "What Does Wilderness Sound Like?" *High Country
News*, November 9, 2018. https://www.hcn.org/articles/photos
-what-does-wilderness-sound-like-alaska.

## Further Reading

Aron, Elaine. "About High Sensitivity, Autism, and Neurodiversity." The Highly Sensitive Person, April 26, 2017. https://hsperson .com/about-high-sensitivity-autism-and-neurodiversity/.

Kish, Daniel. "How I Use Sonar to Navigate the World." TED, 2015. https://www.ted.com/talks/daniel_kish_how_i_use _sonar_to_navigate_the_world/discussion?langua.

McNamara, Shannon. "Coming Out as Human." FemInEm, January 7, 2019. https://feminem.org/2019/01/07/coming-out -as-human/.

Metz, Rachel. "The Smartphone App That Can Tell You're Depressed Before You Know It Yourself." MIT Technology Review, October 15, 2018. https://www.technologyreview.com /s/612266/the-smartphone-app-that-can-tell-youre-depressed -before-you-know-it-yourself.

1440 Multiversity. "Highly Sensitive and Neurodivergent: Nervous System Healing for All." https://1440.org/programs/self -discovery/highly-sensitive-and-neurodivergent/.

Sacks, Alexandra. "A New Way to Think About Motherhood." TED, May 2018. https://www.ted.com/talks/alexandra _sacks_a_new_way_to_think_about_the_transition_to _motherhood?language=en.

Saks, Elyn. *The Center Cannot Hold: My Journey Through Madness*. New York: Hachette, 2007.

Scott, Louise. "Sensory Backpacks to Help People with Autism at Fringe." STV News, July 27, 2018. https://stv.tv/amp/1424895 -sensory-backpacks-to-help-people-with-autism-at-fringe.

# Notes

## Introduction

2   *a slew of articles*: Maria Yagoda, "ADHD Is Different for Women," *The Atlantic*, April 3, 2013, https://www.theatlantic.com/health/archive /2013/04/adhd-is-different-for-women/381158/; Apoorva Mandavilli, "The Lost Girls," *Spectrum*, October 19, 2015, https://www.spectrumnews.org /features/deep-dive/the-lost-girls/.

7   *An entire demographic of women . . .* : Jenny Anderson, "Decades of Failing to Recognize ADHD in Girls Has Created a 'Lost Generation' of Women," *Quartz*, January 19, 2016, https://qz.com/592364/decades -of-failing-to-recognize-adhd-in-girls-has-created-a-lost-generation-of -women/.

## Chapter 1: The Female Mind Throughout History

26   *"dual images of female insanity . . ."*: Elaine Showalter, *The Female Malady: Women, Madness and English Culture, 1830–1890* (New York: Pantheon, 1986), p. 3.

26   *"Biographies and letters . . ."*: Showalter, *Female Malady*, p. 4.

27  *As physicians gained control of the asylums . . .* : Robert Whitaker, *Mad in America: Bad Science, Bad Medicine, and the Enduring Mistreatment of the Mentally Ill* (Cambridge: Basic Books, 2002), p. 29.

28  *"The English have long regarded . . ."*: Showalter, *Female Malady*, p. 7.

28  *"The everyday psychopathology of the masses . . ."*: Gary Greenberg, *The Book of Woe: The DSM and the Unmaking of Psychiatry* (New York: Penguin, 2013), p. 17.

29  *"Surely the doctors who insisted . . ."*: Greenberg, *Book of Woe*, p. 7.

30  *"What seemed never to be in doubt . . ."*: Greenberg, *Book of Woe*, pp. 63–64.

30  *"nearly [doubled] the number . . ."*: Greenberg, *Book of Woe*, p. 40.

31  *"modern female psychology . . ."*: Phyllis Chesler, *Women and Madness: When Is a Woman Mad and Who Is It Who Decides?* (New York: Doubleday, 1972), p. 263.

31  *"Many intrinsically valuable female traits . . ."*: Chesler, *Women and Madness*, p. 263.

31  *"our inner lives are too important . . ."*: Greenberg, *Book of Woe*, p. 8.

32  *"If the people . . ."*: Greenberg, *Book of Woe*, p. 21.

## Chapter 2: Reframing Sensitivity

37  *"Having a sensitive nervous system . . ."*: Elaine Aron, *The Highly Sensitive Person: How to Thrive When the World Overwhelms You* (New York: Citadel, 1996), p. xiii.

37  *"What seems ordinary . . ."*: Aron, *Highly Sensitive Person*, p. 4.

40  *According to the World Health Organization . . .* : World Health Organization, "Mental Disorders Affect One in Four People," October 4, 2001, https://www.who.int/whr/2001/media_centre/press_release/en/.

42  *"While their numbers have increased . . ."*: Maya Dusenbery, *Doing Harm: The Truth About How Bad Medicine and Lazy Science Leave Women Dismissed, Misdiagnosed, and Sick* (New York: HarperOne, 2018), p. 3.

42  *"Until around 1990 . . ."*: Angela Saini, *Inferior: How Science Got Women Wrong—and the New Research That's Rewriting the Story* (Boston: Beacon, 2017), p. 43.

49  *"it is difficult to appreciate . . ."*: Howard C. Hughes, *Sensory Exotica: A World Beyond Human Experience* (Cambridge: MIT Press, 2001), p. 7.

49  *"relies on reflections . . ."*: Hughes, *Sensory Exotica*, p. 9.

49  *"Sometimes it is our own . . ."*: Hughes, *Sensory Exotica*, p. 10.

## Chapter 3: Autism, Synesthesia, and ADHD

56 *"a springboard for discussion . . . "*: Samantha Craft, "Females and As-pergers: A Checklist," The Art of Autism, August 14, 2018, https://the-art -of-autism.com/females-and-aspergers-a-checklist/.

66 *one woman's experience with synesthesia . . .*: Hanna Rosin, Alix Spiegel, and Lulu Miller, "Entanglement," *Invisibilia*, January 30, 2015, https:// www.npr.org/programs/invisibilia/382451600/entanglement.

66 *Synesthesia is the topic of many studies*: Michael J. Banissy, Lúcia Garrido, Flor Kusnir, Bradley Duchaine, Vincent Walsh, and Jamie Ward, "Su-perior Facial Expression, But Not Identity Recognition, in Mirror-Touch Synesthesia," *Journal of Neuroscience* 31, no. 5 (2011): 1820–1824, doi: 10.1523/jneurosci.5759-09.2011; Michael J. Banissy and Jamie Ward, "Mirror-Touch Synesthesia Is Linked with Empathy," *Nature Neuroscience* 10, no. 7 (2007): 815–816, doi: 10.1038/nn1926; L. Maister, E. Tsiak-kas, and M. Tsakiris, "I Feel Your Fear: Shared Touch Between Faces Facilitates Recognition of Fearful Facial Expressions," *Emotion* 13, no. 1 (2013): 7–13.

68 *"The human brain contains . . ."*: Marco Iacoboni, *Mirroring People: The New Science of How We Connect with Others* (New York: Farrar, Straus, and Giroux, 2008), p. 9.

68 *"Cells in the monkey brain . . ."*: Iacoboni, *Mirroring People*, p. 11.

74 *2006 report from the National Autistic Society*: J. K. Kern, M. H. Trivedi, C. R. Garver, B. D. Grannemann, A. A. Andrews, J. S. Savla, D. G. Johnson, J. A. Mehta, and J. L. Schroeder, "The Pattern of Sensory Processing Abnormalities in Autism," *Autism* 10, no. 5 (2006): 480–494, doi: 10.1177/1362361306066564.

75 *2014 study in the* American Journal of Psychiatry: P. Shaw, A. String-aris, J. Nigg, and E. Leibenluft, "Emotion Dysregulation in Attention Deficit Hyperactivity Disorder," *American Journal of Psychiatry* 171, no. 3 (2014): 276–293.

## Chapter 5: Well-Being

118 *"the harmony and order of the universe . . ."*: Duane P. Schultz and Sydney Ellen Schultz, *A History of Modern Psychology*, 11th ed. (Boston: Cengage, 2015), p. 22.

118 *"proposed that the arousal . . ."*: Schultz and Schultz, *History of Modern Psychology*, p. 22.

139 *the role of noise pollution . . .*: Susan Mayor, "Noise Pollution: WHO

Sets Limits on Exposure to Minimise Adverse Health Effects," *BMJ* 2018, 363:k4264; World Health Organization, "Environmental Noise Guidelines for the European Region (2018)," http://www.euro.who.int/en /publications/abstracts/environmental-noise-guidelines-for-the-european -region-2018; Nina Avramova, "Noise: The Other Pollution Hurting Our Health," CNN, October 9, 2018, https://www.cnn.com/2018/10/09 /health/who-noise-guidelines-intl/index.html.

## Chapter 6: Home

146 *public sensory design exhibit . . .* : Michael Kimmelman, "At This Museum Show, You're Encouraged to Follow Your Nose," *New York Times*, April 19, 2018, https://www.nytimes.com/2018/04/19/arts/design/the -senses-review-cooper-hewitt.html.

149 *thinking about design, well-being, and autistic clients*: "Design as Therapy: A Whole New Approach," Healthy Building Science, September 28, 2016, https://healthybuildingscience.com/2016/09/28/design-as-therapy -a-whole-new-approach/.

## Chapter 7: Work

167 *Things are starting to shift . . .* : Quiet Revolution: Unlocking the Power of Introverts, https://www.quietrev.com; SAP, Diversity and Inclusion, "Differently Abled People," https://www.sap.com/corporate/en/company /diversity/differently-abled.html; Microsoft, "Our Inclusive Hiring Programs," https://www.microsoft.com/en-us/diversity/inside-microsoft/cross -disability/hiring.aspx.

177 *"mental health allies . . ."*: Barbara Harvey, "What Companies Can Do to Help Employees Address Mental Health Issues," *Harvard Business Review*, December 18, 2018, https://hbr.org/2018/12/what-companies -can-do-to-help-employees-address-mental-health-issues.

178 *powerful and tender article . . .* : Sarah Kurchak, "The Stories We Don't Tell: My Mom on Raising an Autistic Child and Why She'll Never Write About Me," Medium, March 6, 2018, https://medium.com/@sarah kurchak/the-stories-we-dont-tell-my-mom-on-raising-an-autistic-child -and-why-she-ll-never-write-about-me-79ca1d688626.

## Conclusion

190 *"If teams that set research directions . . ."*: Rediet Abebe, "Why AI Needs to Reflect Society," *Forbes*, November 29, 2018, https://www

.forbes.com/siteds/insights-intelai/2018/11/29/why-ai-needs-to-reflect
-society.

191 *women of color in AI ethics*: Mia Dand, "100 Brilliant Women in AI
Ethics to Follow in 2019 and Beyond," *Becoming Human: Artificial Intelligence Magazine*, October 29, 2018, https://becominghuman.ai/100-brilliant
-women-in-ai-ethics-to-follow-in-2019-and-beyond-92f467aa6232.

191 *"The brain is surprisingly adept . . . "*: Matthew Hutson, "Beyond the
Five Senses," *The Atlantic*, July/August 2017. https://www.theatlantic
.com/magazine/archive/2017/07/beyond-the-five-senses/528699/.

192 *several designers and researchers have written . . .* : Ann Sussman and
Katie Chen, "The Mental Disorders That Gave Us Modern Architecture,"
Common Edge, August 22, 2017, http://commonedge.org/the-mental
-disorders-that-gave-us-modern-architecture/; Stuart Shell, "Why Buildings for Autistic People Are Better for Everyone," Forte Building Science,
https://network.aia.org/HigherLogic/System/DownloadDocumentFile
.ashx?DocumentFileKey=3fff74f0-6418-8e5f-00ed-4ebeb38eabd8&
forceDialog=0.

192 *"good autism-friendly design . . ."*: Kim Velsey, "Autism Informed the
Entire Design of This Revolutionary Boarding School," *Architectural
Digest*, April 3, 2018, https://www.architecturaldigest.com/story/shrub
-oak-international-school.

193 *"these elements help to create . . ."*: Tara Drinks, "NBA Creating Sensory Rooms at More Than Half Its Arenas," Understood, May 4, 2018,
https://www.understood.org/en/community-events/blogs/in-the-news
/2018/05/04/nba-creating-sensory-rooms-at-over-half-of-its-arenas.

211 *There are now hospitals in Canada . . . virtual reality programs . . .
therapists acting as first responders*: "Montreal Museum Partners with Doctors to 'Prescribe' Art," BBC News, October 26, 2018, https://www.bbc
.com/news/world-us-canada-45972348; George Musser, "How Virtual Reality Is Transforming Autism Studies," Spectrum, October 24,
2018, https://www.spectrumnews.org/features/deep-dive/virtual-reality
-transforming-autism-studies/; Zusha Elinson, "When Mental-Health Experts, Not Police, Are the First Responders," *Wall Street Journal*, November 24,
2018, https://www.wsj.com/articles/when-mental-health-experts-not-police
-are-the-first-responders-1543071600.

# Index

# Index

# Index

# Index